Joining of Ceramics

Joining of Ceramics

Edited by

M.G. Nicholas, BSc, PhD, FIM, C Eng

Surface Science and Technology Department
Harwell Laboratory, UK

C, 1990

Published on behalf of

THE INSTITUTE OF CERAMICS
by

CHAPMAN AND HALL

LONDON • NEW YORK • TOKYO • MELBOURNE • MADRAS

UK Chapman and Hall, 11 New Fetter Lane, London EC4P 4EE
USA Chapman and Hall, 29 West 35th Street, New York NY10001
JAPAN Chapman and Hall Japan, Thomson Publishing Japan, Hirakawacho Nemoto Building, 7F, 1–7–11 Hirakawa-cho, Chiyoda-ku, Tokyo 102
AUSTRALIA Chapman and Hall Australia, Thomas Nelson Australia, 480 La Trobe Street, PO Box 4725, Melbourne 3000
INDIA Chapman and Hall India, R. Seshadri, 32 Second Main Road, CIT East, Madras 600 035

First edition 1990

Typeset in 10/12 Times by
Best-set Typesetter Ltd, Hong Kong
Printed in Great Britain at
The University Press, Cambridge

ISBN 0 412 36750 5

British Library Cataloguing in Publication Data
Nicholas, M.G.
Joining of ceramics.
1. Materials. Ceramics. Interfaces
I. Title
666
ISBN 0-412-36750-5

Library of Congress Cataloging-in-Publication Data
Joining of ceramics / [edited by] M.G. Nicholas.
p. cm.
Includes bibliographical references.
ISBN 0-412-36750-5
1. Ceramic materials—Bonding. I. Nicholas, M.G.
TP815.J65 1990
666—dc20

Contents

Advanced Ceramics Reviews *ix*
Contributors *x*
Symbols *xv*
Preface *xiv*

1 OVERVIEW **1**
M.G. Nicholas

2 GLASS–METAL SEALS **7**
A.P. Tomsia and J.A. Pask
2.1 Introduction 7
2.2 Classification of glass–metal seals 8
2.3 Physical factors in making glass–metal seals 11
2.4 Chemical factors in making glass–metal seals 14
2.5 The making of glass–metal seals 26
2.6 Conclusions 28
References 29

3 JOINING GLASS–CERAMICS TO METALS **31**
G. Partridge
3.1 Introduction 31
3.2 The glass–ceramic process 32
3.3 Bonding glass–ceramics to metals 35
3.4 Bond development 49
3.5 Bond quality 52
3.6 Conclusions 54
Acknowledgements 54
References 54

4 METALLIZING CERAMIC SURFACES **56**
A.W. Hey
4.1 Introduction 56

4.2 The ceramic 58
4.3 The primary metallizing layer 59
4.4 The secondary metallizing layer 63
4.5 The brazing stage 65
4.6 Thermal contraction faults 66
4.7 Testing metallized components 69
4.8 Conclusions 71
References 72

5 ACTIVE METAL BRAZING **73**
M.G. Nicholas
5.1 Introduction 73
5.2 Brazes 74
5.3 Brazing 82
5.4 Brazements 85
5.5 Conclusion 92
References 92

6 DIFFUSION BONDING **94**
B. Derby
6.1 Introduction 94
6.2 Mechanisms 96
6.3 The use of interlayers 104
6.4 Bond strength and toughness 105
6.5 Conclusion 110
References 110

7 PHYSICOCHEMICAL ASPECTS OF CERAMIC–METAL JOINING **113**
J.T. Klomp
7.1 Introduction 113
7.2 Interface formation 113
7.3 Interfacial bonding 115
7.4 Joint stability 126
7.5 Conclusions 126
References 127

8 BOND STRENGTH OF CERAMIC–METAL JÓINTS **128**
G. Elssner
8.1 Introduction and general remarks 128
8.2 Testing methods 129
8.3 Bond strength measurements 134
8.4 Concluding remarks 145

Appendix: Analytical interpretation of fracture mechanics bend tests 147
References 153

9 THE NON-DESTRUCTIVE EVALUATION OF CERAMIC MATERIALS AND JOINTS 155
H. Reiter
9.1 Introduction 155
9.2 Ultrasonic testing 155
9.3 Radiography 162
9.4 Thermography and thermal wave interferometry 164
9.5 Concluding remarks 170
References 170

10 JOINING SILICON NITRIDE TO METALS AND TO ITSELF 173
K. Suganuma
10.1 Introduction 173
10.2 Interfacial reactions and microstructures formed between silicon nitride and metals 175
10.3 Residual (thermal) stresses in silicon nitride-metal joints 182
10.4 Other important features 188
10.5 Joining of silicon nitride to itself 190
10.6 Summary 191
References 192

11 CERAMIC–METAL DIFFUSION BONDING – A CASE HISTORY 194
T. Joy
11.1 Introduction 194
11.2 Design objectives 194
11.3 Primary bond requirements 199
11.4 Main steps in the development programme 199
11.5 Process development 200
11.6 Initial problems 202
11.7 The MK I tube 205
11.8 The production run 205
11.9 Conclusions 210
Acknowledgements 211
References 211

Index *212*

Advanced Ceramics Reviews

Series editor:
R. Freer
University of Manchester/UMIST

During the last 15 years there have been major innovative developments in technical and engineering ceramics. Novel techniques for the preparation of powders and ceramics have led to the introduction of new formulations and systems, and significant improvements in the properties of existing materials. The growth of interest in advanced ceramics is reflected in the increasing number of published papers and the establishment of regular specialist symposia. The proceedings from such meetings are usually dominated by contributed papers, and whilst these form a valuable 'snapshot' of research activities at a particular time they are rarely able to give a balanced overall view. The aim of the present series is therefore to provide structured reviews of important subjects within the field of Advanced Ceramics. Each volume will be self-contained, and will consist of a limited number of invited papers. By adopting a broad based approach it is hoped that each review will provide an adequate introduction for the non-specialist, and also act as a reference source, and serve the needs of established workers in materials and related fields.

Contributors

B. Derby
Department of Metallurgy and Science of Materials
University of Oxford
Oxford
UK

G. Elssner
Institut des Werkstoffwissenschaften
Max Planck Institut
Stuttgart
Federal Republic of Germany

A.W. Hey
Morgan Matroc
Stourport on Severn
UK

T. Joy
SERC Daresbury
Warrington
UK

J.T. Klomp
Department of Metals Science and Technology
T.H. Eindhoven
Eindhoven
The Netherlands

M.G. Nicholas
Harwell Laboratory
Oxfordshire
UK

G. Partridge
GEC Alsthom Ltd
Rugby
UK

J.A. Pask
Department of Materials Science and Mineral Engineering
University of California, Berkeley
California
USA

H. Reiter
Department of Materials
University of Bath
Bath
UK

K. Suganuma
National Defence Academy
Yokosuka
Japan

A.P. Tomsia
Department of Materials Science and Mineral Engineering
University of California, Berkeley
California
USA

Symbols

A	interface surface or crack area
A_{ap}	projected bonded area
A_p	projected interface area
A'	absorption coefficient for sound
B	bend test sample dimension in Fig. 8.5
C	constant in equation (2.1)
C_H	specific heat
C'	compliance of bend test sample
C^*	normalized compliance term in equation (8A.13)
C'_0	compliance of unnotched bend test sample
D	grain or pore diameter
D_v	vacancy diffusion coefficient
E	elastic modulus
E_c	elastic modulus of a ceramic
E^*	effective elastic modulus of an interface
F	applied force
F_c	fracture load
G	strain energy release rate
G_c	interface fracture energy
G_F	free energy
G_{Ic}	plane strain fracture energy
I	intensity of a sound wave
I_0	intensity of an incident sound wave
I^*	intensity of an X-ray beam
I_0^*	intensity of an incident X-ray beam
H	enthalpy
ΔH°	enthalpy change
K	stress intensity factor
K_c	interface fracture resistance
K_I	mode I stress intensity factor
K_{II}	mode II stress intensity factor
K_{Ic}	fracture resistance of a monolistic ceramic
L	free surface–vacancy source distance
L'	thickness of disc

P	pore closure stress
P_H	fractional bonded area
R	gas constant
R_F	fraction of thermal waves reflected at interface
S	entropy
S^*	shear modulus
T	temperature
T_0	setting temperature for a glass
T_R	use temperature of a glass
U_E	internal energy of a bend test sample
V	velocity of a sound wave
V_c	velocity of a longitudinal sound wave
V_p	pore volume
V_s	velocity of a shear sound wave
W	bend sample test dimension in Fig. 8.5
W_{ad}	thermodynamic work of adhesion of an interface
W_D	work dissipated during crack propagation
W_s	work due to surface creation
[X]	solute of X
Y_{bi}, Y_G, Y_{iso}, Y_k, Y_0	bend test correction factors defined in equations (8A.14)–(8A.22)
Z	acoustic impedance
a	notch depth in Fig. 8.5
a(X), a(BQ)	chemical activity of an element X or compound BQ
a'	constant in equation (9.5)
a'	Dundurs parameter
a^*	sound attenuation
a^{**}	coefficient of X-ray linear absorption
a_R	coefficient of interfacial sound reflection wave
a_T	coefficient of interfacial sound transmission
b'	constant in equation (9.5)
b^*	constant in equation (9.13)
c	propagation velocity of a sound wave
d	bend test sample dimension in Fig. 8.5
e	bend test sample dimension in Fig. 8.5
f	frequency of a sound wave
h	bend test sample dimension in Fig. 8.5
k	Boltzmann's constant
k	thermal conductivity
k''	constant in equation (8A.5)
ℓ	diffusion path length
ℓ'	thermal diffusivity length
r	radius
t	time

t^*	time for thermal pulse to heat sample to half the maximum temperature
x	distance into sample
x_F	atom fraction
$\Delta\alpha$	difference in coefficients of thermal expansion or contraction
ΔG_F°	free energy change
$\Delta G_{F(XQ)}^\circ$	free energy of formation of compound XQ
ΔG_R	free energy of a reaction
ΔG_R°	free energy of a reaction whose components are in their standard states
ΔH°	enthalpy change
ΔS°	entropy change
ΔT	temperature difference
Γ	amplitude of a thermal wave
Γ_0	initial amplitude of a thermal wave
Ω	vacancy volume
α	coefficient of thermal expansion or contraction
β'	Dundurs parameter
γ_c	surface energy of a ceramic
γ_{cm}	energy of a ceramic–metal interface
γ_{LV}	surface energy of a liquid
γ_m	surface energy of a metal
γ_{SL}	energy of a solid–liquid interface
γ_{SV}	surface energy of a solid
δ	displacement of cracked bend test piece
δ^*	thermal diffusivity
δ_{bf}	failure displacement of a cracked bend test piece
ε	bimaterial constant used in equation (8A.17)
$\varkappa$	constant in equation (8A.7)
λ	sound wavelength
μ	Poisson's ratio
μ^1	chemical potential
μ_c	Poisson's ratio for a ceramic
ρ	density
σ	stress
σ_{bf}	flexural bond strength of a sandwich bend test sample
σ_r	four-point bend rupture strength
σ_R	residual stress
σ_y	tensile stress at crack tip
τ_{xy}	shear stress at crack tip
θ	contact angle between a liquid front and the solid over which it is advancing
ω	angular frequency of a thermal wave

Preface

This book aims to provide the graduate engineer or scientist with an appreciation of the potentials and problems of a range of ceramic joining technologies. Ceramics are growing rapidly in potential as structural and electrical materials for advanced engineering concepts, but reliable joining technologies are vital if these potentials are to be realized; as when complex ceramic structures have to be built up from simply shaped components, when complex structures incorporate more than one ceramic, or when a high performance ceramic is introduced as an insert to a metallic structure.

The topics covered by this book include joining techniques, joint characteristics, fundamental scientific aspects of interface creation and specific descriptions of successful applications. Individual topics are treated in separate chapters and the geographic spread of their authorship is a reflection of the widely recognized need for reliable joining technologies. Some of these technologies are well established but others are barely out of the laboratory, and a number of joint characterization techniques are still in experimental stages. Thus the style of the chapters and the type of information they offer varies.

This book will not make its readers into experts, but it should provide them with sufficient understanding to talk to experts and to select and apply joining technologies in an efficient manner.

M G Nicholas
Harwell Laboratory
UK

1

Overview

M.G. Nicholas

Ceramics and metals are two of the oldest established classes of technologically useful materials. While metals dominate engineering applications, ceramics are widely used to provide thermal and electrical insulation and are emerging as important structural materials. Properties of individual ceramics and metals can vary widely, but the characteristics of most materials in the two classes differ significantly. As shown in Table 1.1 [1–4] ceramics are generally stronger, lighter and more refractory than metals, but are less thermally and electrically conductive and have lower coefficients of expansion. Ceramics have similar stiffnesses to metals but are usually less chemically reactive, and, distinctively, some are transparent. While these generalizations provide good guidance, there are notable exceptions such as the high thermal conductivity of BeO and the refractoriness of molybdenum, mobium and tungsten.

The properties summarized in Table 1.1 reflect the different lattice binding characteristics of metals and ceramics. Metals can be regarded as arrays of positively charged ions set in a continuum of mobile delocalized bonding electrons, whereas ceramics are ionic or covalent materials with bonding electrons whose mobilities are restricted to the immediate locality of one or two ions or atoms. Thus, compared with metals, ceramics are poor electrical conductors and are brittle.

However, ceramics are neither purely ionic nor purely covalent, and according to Pauling [5] the relative degree of ionicity or covalency of ceramic can be related to the electronegativity of its components. From this it follows that in general ceramics become more ionic and less covalent in the sequence borides → carbides → nitrides → oxides. Further, Pauling regarded covalent bonding as intermediate between ionic and metallic bonding, and this can have important implications when the ease with which various ceramics can be joined to metals is considered.

In meeting performance requirements, systems with the characteristics of both ceramics and metals may be required. Examples include view ports in metal structures, insulated electrical conductors, metal devices with

Table 1.1 Physical and mechanical properties of some ceramics and metals

Material	*Melting temperature (°C)*	*Density ($Mg\ m^{-3}$)*	*Thermal expansion coefficient ($\times 10^{-6}\ K^{-1}$)*	*Strength** *(MPa)*	*Elasticity (GPa)*	*Thermal conductivity ($W\ m^{-1}\ K^{-1}$)*	*Electrical conductivity ($V^{-1}\ m^{-1}$)*
HfB_2	3200	11.2	7.6	98.5		26	11×10^6
B_4C	2350	2.5	4.3	352	457	26	2.6×10^3
SiC	2700†	3.2	4.3	141	211	50	1×10^3
TiC	3140	4.9	7.2	703	422	36	1.9×10^6
WC	2777	15.8	5.2	598	704		
Si_3N_4	1900†	3.2	2.5	211	176	17	1.6×10^{-9}
TiN	2900	5.4	8.1			17	
Al_2O_3	2050	4.0	7.9	457	387	35	10^{-14}
BeO	2530	3.1	7.4	246	401	210	1.1×10^{-10}
MgO	2800†	3.6	11.6	281	394	62	10^{-12}
SiO_2	1710	2.3	3.0‡	88	354	1.5	10^{-12}
ZrO_2	2960	5.6	7.5	176	140	19	3×10^{-5}
Ag	962	10.5	19.1	172	83	128	2.1×10^7
Al	660	2.7	23.5	55	71	238	3.7×10^7
Cu	1083	9.0	17.0	216	130	397	5.9×10^7
Fe	1535	7.9	12.1		211	78	1×10^7
Mo	2615	10.2	5.1	435	325	137	1.7×10^7
Nb	2467	8.6	7.2	240	105	54	0.6×10^7
Ni	1455	8.9	13.3	310	199	89	1.4×10^7
Sn	232	7.3	23.5		50	73	0.8×10^7
Ti	1677	4.5	8.9	241	120	22	0.2×10^7
W	3387	19.3	4.5	550	411	174	2×10^7

*Four-point bend test data for ceramics; tensile data for metals.
†Does not melt, but sublimes, decomposes or vaporizes.
‡Crystalline SiO_2; the coefficient for vitreous SiO_2 is $0.5 \times 10^{-6}\ K^{-1}$.

wear-resistant surfaces and ceramic bodies with exceptional toughness and tensile strength. These and many other applications where it is necessary or at least desirable to use ceramics and metals in combination call for macroscopic joints to be formed between materials which are sometimes very dissimilar.

This book is concerned with the joining together of macroscopic ceramic and metal workpieces in combinations that are unknown in nature. However, there are many natural examples of microscopic combinations, such as the bonding of carbide precipitates in metals or the oxide films present on their surfaces. Theories and models advanced to explain the bonding of these natural systems have also been of great value in the development of synthetic microscopic combinations such as ceramic fibre reinforcement of metal matrices or the screen printing of microcircuitry onto ceramic substrates. More generally, models developed and information gathered from the study of both natural and synthetic microscopic combinations are often also relevant to macroscopic systems. Thus in attempting to produce strong bonds it is advantageous if the electron structures of the workpieces are similar – so it is generally easier to bond metals to borides or carbides than to nitrides or oxides – and if the interfaces are in chemical equilibrium and oriented so as to minimize the misfit of the workpiece lattices.

In principle many techniques could be used to join ceramics to metals, ranging from welding to mechanical attachment. The possibilities are too numerous to cover in a book of this size and, more importantly, not all are relevant to the joining of macroscopic structural ceramic components.

Welding can be regarded as the ultimate technique for producing high-integrity joints, and some workers have achieved technical success. Thus Hokanson [6] produced electrical feedthroughs by electron beam welding wires of molybdenum, mobium and tungsten into Al_2O_3 insulators, and Rice [7] has described the use of electron beams and arcs to form Al_2O_3–Ta, C–Ta, ZrO_2–Mo, ZrO_2–Nb and ZrO_2–Ta couples with strengths of up to 210 MPa. However, the applicability of welding to ceramic–metal systems is very limited. For successful welding, the melting points and thermal contraction characteristics of the workpieces should be closely matched and this will rarely be achievable, as indicated by the data in Table 1.1. Further, the thermal contraction characteristics of the material formed in the weld pool must match those of the metals. Difficulties can also arise if the ceramics sublime or decompose rather than melt, or if either the ceramic or the metal undergoes phase changes on cooling. Thus the problems of identifying potentially weldable systems are substantial, and welding is impractical at present for the considerable majority of systems of technical interest.

In contrast, mechanical attachment of metals to ceramics is readily achievable using press-fitting and shrink-fitting. The production of light duty joints such as those in domestic fuses is substantial, and there are also well-established heavy duty industrial applications – for example, the use of hooks or dogbones to support furnace roof refractories. Recent high-technology applications include the clamping of ceramics to the leading edges of the NASA space shuttle [8] and the bolting and shrink-fitting of ceramic inserts for high-temperature high-performance reciprocating engines [9]. However, mechanical attachment is not a generally attractive process for joining structural ceramics because of the frequent necessity for introducing intrusive stress-raising features such as threads or bolt holes, and the low strengths and lack of hermeticity of the joints. A direct bond between the workpieces is required for many applications.

It can be argued that the most appropriate direct bonding techniques are glazing, brazing and diffusion bonding as described in later chapters. Thus the initial designs for mechanical attachment of ceramic inserts in high-performance engines are being replaced by brazed joints. The advantage of these techniques over welding and some mechanical attachment processes is that they leave the workpieces substantially unaffected by the joining process. This is also true of adhesive bonding, but this technique has not been included for detailed description in this book because many of the potential applications for ceramic–metal systems require the joints to withstand high stress, high temperatures or chemically arduous environments, conditions unsuitable for adhesives used in commercial practice at present.

Glazing, brazing and diffusion bonding are described separately in the following five chapters. The specific requirements for successful application can be very disparate, but nevertheless the processes display some common features. Thus glazing and brazing both depend on a liquid bonding phase wetting the workpieces and can be regarded as mirror image processes. While a glaze is a ceramic and hence the joint interface experiencing the most marked bonding discontinuity is that formed with the metal workpiece, the metallic nature of brazes makes the critical interface that formed with the ceramic workpiece. It is usually good practice to oxidize the surfaces of metal workpieces before they are glazed (Chapter 2 and 3) and metallization of ceramic workpieces is frequently adopted as a brazing promoter (Chapter 4). In active metal brazing (Chapter 5) ceramic metallization occurs during brazing through the formation of low-valent reaction products. The formation of monovalent oxides on the metal surface also plays a crucial role in glazing. While diffusion bonding differs from glazing and brazing in not involving flow of a liquid phase, the driving force for interface formation, i.e. energy minimization, is the same (Chapter 6). Similarly, workpiece expansion coefficients and the extent of interfacial chemical reactions play important roles in determining the strengths of glazed, brazed and diffusion-bonded joints. When a ductile metal interlayer is used to promote diffusion bonding, superficial similarities to brazed joints are so marked that the process could be described as solid phase brazing.

Whatever the technique employed, the objective remains the creation of high-integrity joints between structural ceramics and metals. Integrity can be measured in various ways, and in this book chapters are devoted to the chemical stability and continuity of interfaces and to the strength of joints (Chapters 7 and 8). As will be illustrated in Chapter 7, the chemical integrity of a particular interface depends not only on the compositions of the workpieces but also on those of the fabrication and service environments. Physical integrity – continuity – should be achieved if the glaze or braze wets the workpieces or if the diffusion-bonding conditions are chosen so as to maximize micropore closure, but verification of continuity at interfaces between dissimilar materials can pose problems and require the use of very specialized techniques (Chapter 9). If unbonded regions are present at the interfaces, they can have a major detrimental effect on the impact and toughness characteristics. Strength parameters can also be degraded if the joint interfaces are chemically unstable or if thick layers of reaction products are formed during the joining process.

It is difficult to optimize the conditions for joint formation and to verify that the product is of the desired quality. Further, the optimum interface bonding conditions may be unacceptable, producing a permanent softening of a metal workpiece for example, and hence joints with diminished integrities may sometimes have to be used. However, these joints can perform perfectly adequately in service. References to actual and potential applications are contained in the chapters concerned with fabrication methods and joint integrity assessments. Discussions of specific applications are also contained in the two final chapters. In Chapter 10 the development of techniques for bonding Si_3N_4, which is emerging as a prime engineering ceramic, is reviewed. Considerable progress has been made, and some of the techniques are now being exploited in demanding applications such as high-performance reciprocating engines. Finally, the book concludes with a major success story in Chapter 11 where the development programme that enabled the Daresbury Laboratory to use diffusion bonding in the fabrication of the accelerator tube for their Nuclear Structure Facility is described. Painstaking optimization and rigorous control over processing enabled the Laboratory to fabricate a world-class facility able to operate continuously under the most demanding conditions.

REFERENCES

1. Lynch, J.F., Ruderer, C.G. and Duckworth, W.H. (1966) *Engineering Properties of Selected Ceramic Materials*, American Ceramic Society, Columbus, OH.

2. Samsonov, G.V. (1964) *Handbook of High Temperature Materials*, No. 2, *Properties Index*, Plenum, New York.
3. Samsonov, G.V. (1982) *The Oxide Handbook*, Plenum, New York.
4. Brandes, E.A. (1982) *Smithells Metals Reference Book*, Butterworths, London.
5. Pauling, L. (1939) *The Nature of the Chemical Bond*, Cornell University Press, Ithaca, NY.
6. Hokanson, A., Rogers, S.L. and Kern, W. (1963) *Ceram. Ind. J.*, **81**, 44.
7. Rice, R.W. (1970) *Welding of Ceramics*, Report 7085, National Research Laboratory Washington, DC.
8. Rice, R.W. (1975) in *Advances in Joining Technology* (ed. J.J. Burke), Brook Hill, Cambridge, MA., p. 69.
9. McLean, A.F. (1986) in *Ceramic Materials and Components for Engines* (eds W. Bunk and H. Hausner), DKG, Bad Honnef., p. 1023.

2

Glass–metal seals

A.P. Tomsia and J.A. Pask

2.1 INTRODUCTION

Glass–metal seals have found wide application in electrical engineering and electronics. Incandescent and vapour lamps, electron tubes and housings for semiconductors are some of the many applications that use glass–metal seals for the combination of either optical or insulating properties and metallic conductivity. In a broad sense the field also includes porcelain enamelling, high-temperature protective coatings, joining ceramics to ceramics or metals through a glassy phase (as in high-pressure sodium lamps, magnetic recorder heads, high-voltage feed-throughs etc.) and composite materials.

Glass is used in composite structures such as lamps and electronic tubes because it transmits light, can easily be formed into useful shapes and provides electrical insulation and chemical inertness in corrosive or oxidative environments. In protective coatings and composites the glass resists oxidation, especially at elevated temperatures. The primary focus of this chapter is glass–metal seals. However, many of the theoretical principles also apply to glass–glass, glass–ceramic and ceramic–metal joining. Glass–metal systems can also be joined by adhesives or by metallizing and soldering techniques, but they are less usual and will not be covered in this chapter.

In most lamp applications, whether the internal atmosphere is high vacuum or pressure, the electrical leads must be hermetically sealed to the glass which requires strong adherence between glass and metal. The basic requirements for strong glass–metal seals are chemical bonding and minimal stress differentials at the interface with favourable stress gradients in the interfacial zone. It is generally recognized that a number of factors must be considered and controlled if a satisfactory bond or seal between two dissimilar constituents is to be produced. These factors can be grouped into three general categories: physical, chemical, and processing.

A vast number of reports and monographs on the general problem of bonding, some specific to the area of glass–metal seals and others dealing with related topics, have been published. The bibliography on this subject, covering the period from 1950–89, contains nearly 2500 citations, even excluding the extensive patent literature pertaining to the topic. The size of the available data base reflects the complexity of the subject, and also its considerable industrial importance. Although interpretations of experimental results have varied, each study has made some contribution towards the development of an improved overall understanding of the conditions which promote bonding and adherence. The term 'bonding' as used in this chapter refers to chemical bonding at the interface, whereas mechanical adherence strength also takes into account the effect of physical factors on the strength of the assembly.

2.2 CLASSIFICATION OF GLASS–METAL SEALS

Glass–metal seals can be classified in a number of ways. The following condensed list is not complete but it provides an indication of the terminology and also reflects the wide variety of approaches that can be used. Seals can be classified and described in the following terms:

1. the geometry of the metal parts (bead, tubular, disc, ribbon, and feather-edge (Housekeeper));
2. the type of metal or alloy used for the metallic parts (molybdenum, platinum, copper, Kovar, Fe–Ni alloys, etc.);
3. the type of glass used (soft glass (soda-lime–silica, lead oxide–mixed alkali oxide–silica, etc. with working temperatures of 800–1000 °C and coefficients of thermal expansion above $5 \times 10^{-6}\ K^{-1}$)) and hard glass (borosilicate and aluminosilicate glasses and vitreous silica with working temperatures of 1000–1300 °C, and coefficients of thermal expansion below $5 \times 10^{-6}\ K^{-1}$));
4. the type of joining (mechanical, adhesive, fused glass, solder glass or metallization);
5. the fabrication technique (flame, induction, frit, pressure-diffusion, tape or high-frequency resistance welding);
6. the level of stress in the glass (matched or balanced, and unmatched or unbalanced (compression and ductile metal or Housekeeper)).

The most frequently used classification is the last. In matched seals, the coefficients of thermal expansion of the glass and metal are similar over the entire temperature range of sealing. Sealing is the result of good chemical bonding between the glass and the metal at the interface. Matched glass seals can be made with metals, ceramics or other glasses. Figure 2.1 shows a typical design of a matched glass–metal seal.

In unmatched seals, the coefficients of thermal expansion of the various parts are mismatched. These seals are divided into two groups: compression seals and ductile metal or Housekeeper seals. In compression seals, sealing

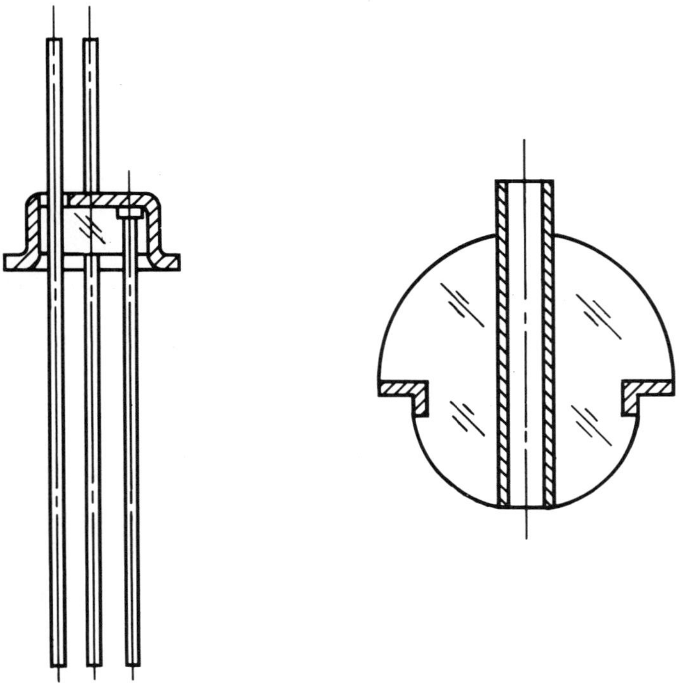

Fig. 2.1 Typical design of matched glass–metal seal.

is accomplished by the metal establishing a compressive hoop force on the glass. With these seals mechanical bonding can be considered sufficient; however, development of chemical bonding is always desirable. Compression glass seals are made with the metals only applying the force. Figure 2.2 shows a typical design of a compression seal.

Ductile metal or Housekeeper seals were developed by G.W Housekeeper [1] to join many types of glasses to copper. Figure 2.3 shows typical steps in making a Housekeeper seal between copper and glass tubing. Copper has a relatively high coefficient thermal expansion ($17 \times 10^{-6}\ K^{-1}$) and does not match any commercial glass in large-scale production. However, because of its ductility, copper can be joined to almost any glass if the Housekeeper seal is used. Because of its high ductility, copper, if it is thin enough, adapts itself to large coefficient of thermal expansion differences with sealed glasses so that no cracking of the seal occurs. Housekeeper seals are now also used with Kovar and several stainless steels, but this type of seal requires good chemical bonding between glass and metal.

Examples of matched glass–metal seals can be divided into two additional groups: seals with hard glasses (low-expansion glasses of the

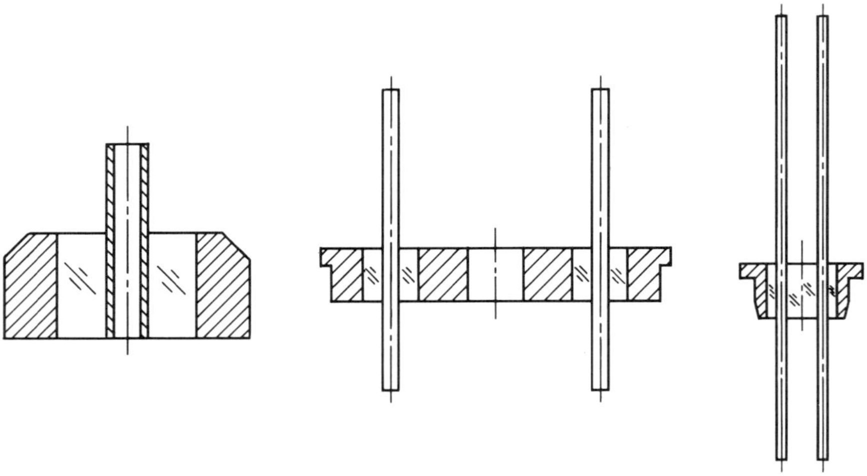

Fig. 2.2 Typical design of compression glass–metal seal.

borosilicate type) and seals with soft glasses (high-expansion glasses of special types).

Compression glass–metal seals can be divided into three subcategories [2]:

1. matched compression seals (standard type) where the glass and pins have matched thermal expansion coefficients but the compression ring (frame) has a considerably higher coefficient of thermal expansion

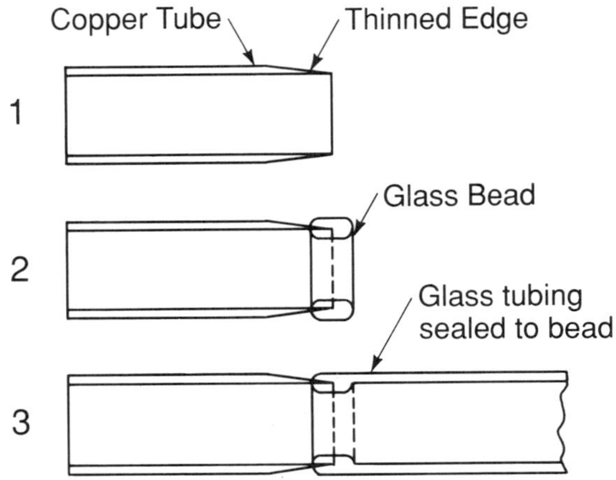

Fig. 2.3 Steps in making a Housekeeper seal between copper and glass tubing: 1, the copper tube is thinned; 2, glass bead is applied to the edge of the thinned copper tube; 3, the glass tubing is then sealed to the bead.

(frame, steel, 13×10^{-6} K^{-1}; glass, soft, 9×10^{-6} K^{-1}; pin, Ni49 [50Ni, 1Cr, 49Fe], 9×10^{-6} K^{-1});

2. low-expansion matched compression seals using hard glass whose coefficients of thermal expansion are lower by about 5×10^{-6} K^{-1} (frame, Ni49, 9×10^{-6} K^{-1}; glass, hard, 5×10^{-6} K^{-1}; pin, Kovar, 5×10^{-6} K^{-1});
3. reinforced compression seals where the thermal expansion of the pin is lower than that of the sealing glass resulting in reinforcement (amplification) of the prestresses on the pin (frame, steel, 13×10^{-6} K^{-1}; glass, soft, 9×10^{-6} K^{-1}; pin, kovar, 5×10^{-6} K^{-1}).

2.3 PHYSICAL FACTORS IN MAKING GLASS–METAL SEALS

The most important requirement in developing a successful glass–metal seal is a proper matching of the thermal expansion and contraction characteristics of the bonded materials. Ideally, one strives to develop a slight compressive stress in the glass component of the seal during cooling from the firing or maturing temperature because the compressive strength of glasses usually exceeds their tensile strength by a factor of about 10–20. Thermal expansion mismatch also results in shear stresses acting along the glass–metal interface, and the bond developed must be of sufficient strength to withstand these stresses, otherwise adherence failure will occur. If bonding is weak, a crack will propagate at the interface, causing separation of the bonded components. If bonding is strong and the difference between the coefficients of thermal expansion is large, a crack will nucleate in the glass very close and parallel to the interface.

In considering thermal expansion/contraction characteristics, the temperature range of interest is more extensive than is sometimes appreciated. In order to obtain a correct indication of the interfacial strain, it is essential to consider thermal expansion behaviour up to the set point temperature of the glass, and not just to 300 °C (which are the data usually given). This point is illustrated in Fig. 2.4 for tungsten sealed to a Corning 772 glass. If a seal is assumed to be stress free at 300 °C, on cooling a differential contraction of 0.000 25 unit/unit would be predicted at room temperature. However, stress development initiates in the seal upon cooling once the temperature drops below the set point temperature or ‘solidification’ point. If the entire thermal expansion curve is evaluated, a much lower (and more acceptable) differential expansion of only 0.000 07 unit/unit results. By examining the offset between the curves representing the metal and the glass, we can see that the actual concern in this particular case would be the integrity of the system *during cooling*, when the system undergoes a maximum (transient) contraction differential of 0.000 32 unit/unit.

Failure of glass–metal and ceramic–metal assemblies is often thought to arise at regions of high residual stress. The transition zone between the

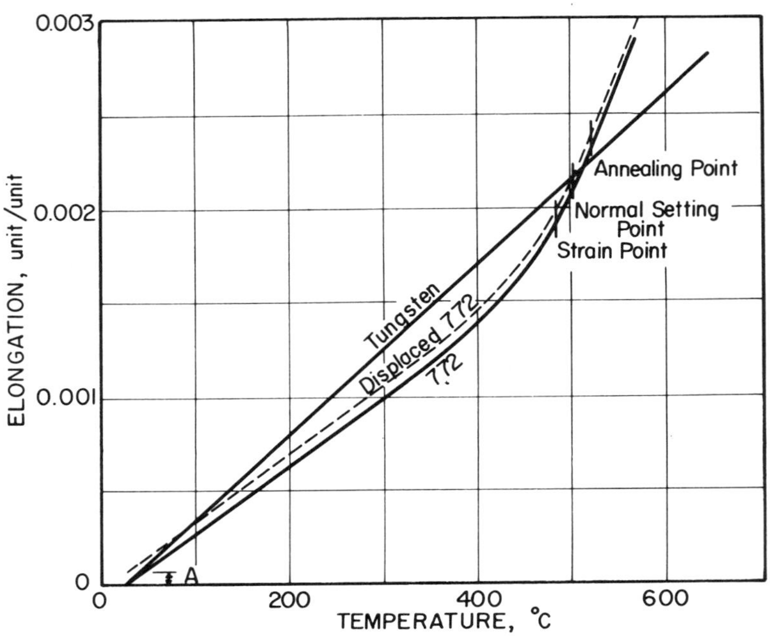

Fig. 2.4 To determine the differential expansion in a glass–metal seal, the glass expansion curve is displaced until the set point for the conditions of cooling coincides with the metal expansion curve. The spread between these curves then indicates the differential expansion, which governs the degree of stress on cooling. The indicated differential expansion *A* determines the internal stresses in the seal at room temperature.

metal and the glass (the interfacial zone) is one region where a high residual stress and flaws can arise. This is mainly due to reactions at the interface and diffusion of the metal ions into the glass. As a result of this process, the properties of this interfacial zone, such as the coefficient of thermal expansion, the glass transition temperature and the elastic modulus, will probably be different from those of the bulk glass. Stress-induced failure in glass–metal systems is a subject receiving much attention, but as yet no method has been satisfactorily developed for the evaluation of stresses in these systems.

Residual stresses in glass–metal systems arise from several sources, but the dominant one is the difference between the thermal expansion coefficients of the metal and the glass, which results in different contractions during cooling from the softening temperature to the use temperature. Glasses do not transform from a liquid to a crystalline solid when cooling but become increasingly viscous. A series of characteristic temperatures, or points, that are of importance to the technologist can be identified. Those of particular note include the following: the working temperature at which the viscosity is 10 N s m^{-2} (10^4 P) and below which the glass cannot be manipulated; the softening temperature below which glass is significantly rigid, as denoted by a viscosity of $10^{4.5} \text{ N s m}^{-2}$; the annealing temperature below which it takes longer than 15 min for the glass to be stress relieved and at which the viscosity is $10^{10} \text{ N s m}^{-2}$; the strain temperature below which it takes longer than 4 h for the glass to be stress relieved and at which the viscosity is $10^{11.6} \text{ N s m}^{-2}$; the setting temperature at which the metal and glass components of a seal are in equilibrium and below which elastic stresses are generated owing to mismatches between the coefficients of thermal expansion (this is not a fixed temperature but generally lies between the annealing and strain temperatures). The magnitude of the residual stress σ_R is given by

$$\sigma_R = CE \int_{T_R}^{T_0} \Delta a \Delta T \qquad (2.1)$$

where Δa is the difference between the coefficients of thermal expansion of the metal and the glass (in general a function of temperature), T_R is the use temperature, T_0 is the setting temperature, E is the elastic modulus and C is a constant which is dependent upon the geometry and relative thicknesses of the two materials. There is some sensitivity of σ_R to thermal history because the set point can be dependent upon cooling rate and thermal cycling [3], and because the thermal contraction of the glass can be influenced by crystallization, at temperatures below T_0, of phases with different densities from that of the parent glass. Such changes in thermal expansion may produce serious thermal expansion mismatch. Thermal gradients in the glass produced by rapid heating and cooling can also induce either permanent residual stresses (if the gradient exists during cooling through the set point, i.e. thermal tempering) or transient stresses (if the gradient is established at temperatures below the set point). Therefore residual stresses can be decreased by control of thermal history. However, for a given glass–metal system, a minimum residual stress level at room temperature is dictated by the expansion coefficient difference Δa and the geometry of the system (through the constant C). The dependence of σ_R on geometry provides a strong incentive for devising methods to measure residual stresses in actual components, rather than in specially constructed test specimens.

Another physical characteristic or factor that is important is the nature of surfaces being bonded. A highly irregular surface finish, produced by either abrasion or chemical erosion, can provide a mechanical contribution to adherence. There is evidence that surface roughness produced by sandblasting or chemical etching of the metal substrate can provide mechanical interlocking and also increase the surface area for glass attachment [4]. Although interfacial roughening contributes to adherence, produces a more diffuse (wider) transition zone, and increases the contact area, it is not necessary if a chemical bond (i.e. a continuous electronic structure) is attained across the interface and the difference between the coefficients of thermal expansion is not too large.

The adherence between dissimilar phases which are brought into contact results from the interaction of surface (and possibly near-surface) atoms, ions and molecules of the individual phases. These interactions stem from a difference between the environments of atoms on the surface and those in the interior and result in a net force of attraction between the contacting surfaces. Atoms in the interior of a solid or liquid are symmetrically coordinated (and bonded), while those on the surface are asymmetrically coordinated. As a result, surface atoms have excess energy compared with bulk atoms. This difference is manifested as the surface energy. The high surface energy of metals (ranging from $0.48\ J\ m^{-2}$ for lead to $1.84\ J\ m^{-2}$ for iron) produces a strong driving force for shape changes that reduce the interfacial area and chemical changes that lower the surface energy; both produce a lower energy state.

The surface energy can be lowered through chemical (surface) interactions. Metal surfaces readily adsorb gases. On an atomic scale, dipole–dipole interactions, electron transfer, covalent bond formation etc., which can be termed chemical or electrical interactions, may take place and result in bonding [5]. Some gases, e.g. oxygen, react chemically to form oxides because of the strong electropositive nature of the metal atoms and the thermodynamic driving force. Even at low pressures and high temperatures, monolayers of oxygen or oxide persist on metal surfaces that appear 'clean', including the surfaces of noble metals. As a result, the high sensitivity of metal surfaces to oxygen, and possibly to other gases as well, must be taken into account when attempting to understand adherence.

2.4 CHEMICAL FACTORS IN MAKING GLASS–METAL SEALS

The two most important chemical steps in the development of good adherence at glass–metal interfaces are (1) formation of an intimate atomic contact interface, followed by (2) reactions to reach stable chemical equilibrium at the interface which result in an electronic structure across the interface, i.e. a chemical bond. An intimate solid–liquid interface is formed when the liquid glass wets or spreads, thereby penetrating irregularities and distributing itself on the metal surface. Reactions invariably form equilibrium phases. At glass–glass and ceramic–ceramic interfaces either solution or compound formation takes place readily to reach the lowest free-energy composition. The simplest reaction is the formation of a solution of one phase in the other, leading to and maintaining equilibrium saturation in both phases at the interface. A continuation of the reaction is associated with diffusion into the bulk. Any compounds that form are chemically compatible with both phases.

However, glass–metal and oxide ceramic–metal interfaces are not compatible, and redox reactions involving oxidation of the metal and reduction of a cation in the glass or ceramic take place at the interface [6,7]. In the case of an oxide phase, the oxygen released by the reduction of the cation forms an oxide with the metal. Saturation of the interface with the metal oxide and subsequent formation of a molecular or multimolecular layer of oxide results in equilibrium and chemical bonding at the interfaces since the oxide layer is compatible with both its metal and the glass or ceramic. Commercially, the necessary compatible oxide layer at the interface is obtained by pre-oxidizing the metal, which is a redox reaction, before the application of the glass [8].

2.4.1 Wetting and spreading

In studying the wetting behaviour of glass–metal systems, high temperatures must be employed so that the glass will be in the liquid state. Wetting occurs when the surface energy of the solid is reduced by contact with the liquid. Liquid glass can wet a metal when either van der Waals bonding or chemical bonding is present at the interface. However, wetting is favoured most by the formation of strong chemical bonds at the interface resulting in a smaller contact angle (θ in Fig. 2.5).

Very few systems are initially in chemical thermodynamic equilibrium. As a result, the possibility of a chemical reaction is almost always present. If the reaction does not occur because of unfavourable sluggish kinetics, then a state of metastable chemical equilibrium and van der Waals bonding exists. If a reaction does occur in response to the driving force towards stable chemical equilibrium, it initiates at the interface. During this transient stage, as equilibrium is approached interfacial energies are modified and the measured contact angles are therefore affected. If the existence of such reactions is not recognized, as may be the case when the reactions are proceeding slowly, the measured and reported values of the contact angles do not correspond to the stable equilibrium state and are in this respect incorrect. Unfortunately, owing to this effect, many of the reported results in the literature are incorrect or suspect. If the reaction proceeds sufficiently rapidly, spreading will occur as will be discussed later.

Extensive studies of wetting behaviour have been made using the sessile drop technique. Such experiments provide a means of studying wetting and spreading of a liquid on a rigid solid and determining relative surface and interfacial energies. Proper interpretation provides information on whether a reaction is taking place at the interface and on the occurrence and effect of contamination of the surfaces. The technique also provides a convenient

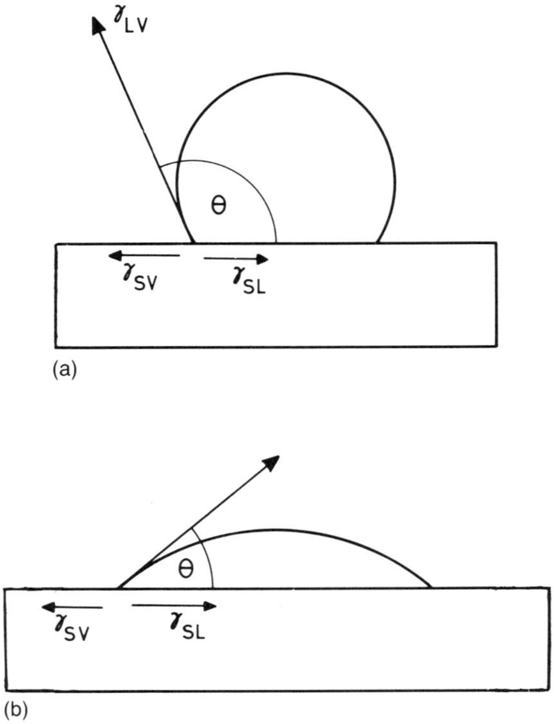

Fig. 2.5 Schematic diagrams of sessile drop configurations: (a) non-wetting drops; (b) wetting drops. The contact angle is identified as θ.

method of studying the reactions that occur at the interface by characterizing cross-sections obtained by cutting a frozen sessile drop assembly perpendicularly to the interface.

Sketches of cross-section planes perpendicular to the interfaces and passing through the vertical axes of both wetting and non-wetting molten sessile drops are shown in Fig. 2.5. In each case the configuration is determined by the solid rigid phase and the surface energies of the phases. The relative order of surface energy values is metals > crystal oxides > oxide glasses. Under non-reactive conditions, wetting represented by a steady state acute contact angle ($\theta < 90\,°C$) always occurs when $\gamma_{SV} > \gamma_{SL}$, and for glass–metal systems γ_{SL} is invariably larger than γ_{LV} and so wetting is associated with the condition $\gamma_{SV} > \gamma_{SL} > \gamma_{LV}$. ($\gamma$ represents a solid or interfacial energy identified by the subscripts SV for the solid surface, SL for the solid–liquid interface, and LV for the liquid surface.) This configuration is typical of a liquid glass drop on a metal substrate. Also under non-reactive conditions, non-wetting, represented by a steady state obtuse contact angle ($\theta > 90\,°C$) always occurs when $\gamma_{SV} < \gamma_{SL}$, even though an intimate interface forms, and typically for a metal drop on a glass or ceramic substrate $\gamma_{LV} > \gamma_{SL}$ and $\gamma_{SV} > \gamma_{SL} > \gamma_{LV}$. Thus γ_{SL} for both wetting and non-wetting glass–metal systems is between γ_{SV} and γ_{LV}, i.e. the higher surface energy is reduced in contact with the lower surface energy phase. Wetting is defined as the reduction of γ_{SV} of the rigid substrate by the liquid. In the case of non-wetting, the surface energy of the rigid substrate is not reduced by the liquid; however, an intimate interface, is formed because the surface energy of the liquid is reduced by the rigid substrate.

The driving force for wetting is thus $\gamma_{SV} - \gamma_{SL}$ which is acting on the periphery of the liquid drop. The resisting force is γ_{LV} since the lowest surface energy configuration for the liquid is a sphere. A balance of these forces gives us a stready state contact angle and the familiar Young equation:

$$\gamma_{SV} - \gamma_{SL} = \gamma_{LV} \cos\theta \tag{2.2}$$

In the absence of a reaction, whether stable or metastable chemical equilibrium exists, the driving force for wetting never exceeds γ_{LV} and spreading or dynamic extension of the liquid drop does not occur [9].

Examples of the value of understanding the significance of these relationships are given below. If γ_{SV} is known to be greater than γ_{LV} but non-wetting is observed, then the experimental γ_{SL} must be greater than both γ_{SV} and γ_{LV}. The conclusion is that a wetted interface between the solid and liquid in question has not formed because of the presence of some unreactive contamination on the solid surface. However, if u $< 90°$, indicating that γ_{SL} is between γ_{SV} and γ_{LV}, a true wetted interface is present. Furthermore, if $\gamma_{LV} > \gamma_{SV}$ and γ_{SL} is between them, a true interface is again present but the non-wetting liquid will not penetrate the surface irregularities.

When a reaction occurs, the free energy of reaction per unit interfacial area and unit time, $d(d\Delta G_R/dA)/dt$, enhances the driving force for wetting to produce the inequality

$$\gamma_{SV} - \left\{\gamma_{SL} + \frac{d(d\Delta G_R/dA)}{dt}\right\} > \gamma_{LV} \cos\theta \tag{2.3}$$

but only if the solid is an *active participant*. In a solution reaction, if only one phase is unsaturated relative to the other, the unsaturated or solvent phase is changing its composition and the solute phase is not. We can then define the solvent as the active participant and the solute as the passive participant. If both phases are unsaturated relative to each other or they form a compound at the interface, then both phases are active participants. If the enhanced driving force for wetting exceeds γ_{LV}, then spreading occurs, which can happen when γ_{SV} is either greater or smaller than γ_{LV} [10]. The dynamic stage of movement or spreading stops when the liquid is completely reacted.

In glass–metal systems, the contact angle u formed by the liquid glass on the metal substrate is acute and the glass surface energy is less than that of the metal. Surface energies of silicate glasses do not vary significantly and are generally in the range 0.25–0.4 J m^{-2}. The surface energies of metals always exceed those of glasses and are typically in the range 0.5–2.0 J m^{-2}. The acute contact angle is the result of wetting of the solid by the liquid. The surface energy of the metal in contact with the liquid glass (γ_{SL}) is less than that for metal in contact with the vapour (γ_{SV}). The reduction in surface energy stemming from coverage by the liquid provides the driving force for wetting. The resistance to continued wetting of the solid by the liquid drop stems from the eventual increase in the total surface energy of the liquid as further extension occurs or by completion of the reaction.

2.4.2 Nature of reactions at glass–metal interfaces

It has been indicated that chemical bonding at glass–metal interfaces is associated with the presence of stable chemical equilibrium. Equilibrium

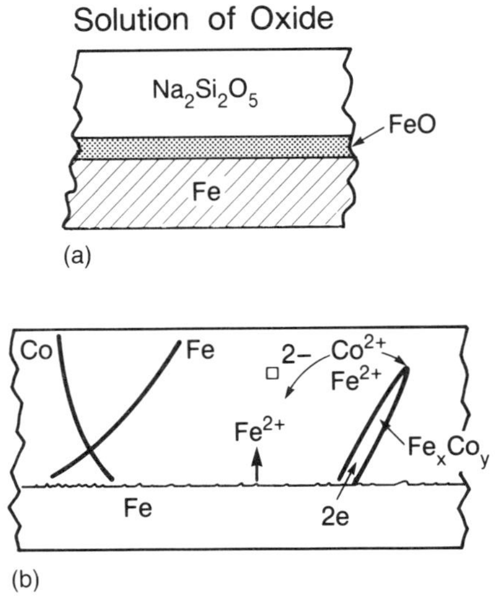

Fig. 2.6 Schematic cross-sections of (a) glass on pre-oxidized metal and (b) the redox reaction of glass containing CoO with the iron substrate to form Fe_xCo_y dendrites by a microgalvanic cell mechanism with an electronic circuit composed of electrons and negative vacancies.

requires the achievement of phase compatibility by saturating the interfacial zone with a low-valent oxide of the substrate metal. A transition layer of the oxide, which theoretically need only be one molecular layer thick, is produced at the interface. This structure or microstructure is realized by redox reactions.

2.4.3 Pre-oxidation of metal

The normal technological procedure for achieving chemical bonding and favourable stress gradients in glass–metal assemblies is to pre-oxidize the metal and apply the glass as shown in Fig. 2.6(a) in which the model system of $Na_2Si_2O_5$ glass and metallic iron is used as an example [11–13]. At temperature, the molten glass wets and dissolves the oxide. The glass at the oxide interface immediately becomes saturated because the solution rate of the oxide is faster than the rate of diffusion of the dissolved oxide into the bulk glass. Chemical bonding at the interface is thus realized. The oxide layer also bonds chemically to the metal which is saturated with the oxide, at least at the surface. The saturated interfaces and the oxide layer thus have a chemical activity of unity for FeO, which is a requirement for chemical bonding in this system.

As the dissolved oxide diffuses into the glass an activity and concentration gradient is formed as shown schematically in Fig. 2.7(a). It can be seen that, as more oxide dissolves, the concentration gradients become more extended, but the equilibrium saturation at the oxide–glass interface and consequently the bonding are always maintained because of the faster reaction rate. Such an addition of a metal oxide to the glass affects its coefficient of expansion because of an increase in the ratio of oxygen to silicon and the introduction of cations with different degrees of covalency. Concentration gradients are proportional to thermal expansion coefficient gradients, and hence more extensive diffusion will generally result in more favourable stress gradients [14]. It can be seen that the activity and concentration gradients become more extended with increasing thicknesses of starting oxide. It is thus evident that many, if not most, experiments based on varying oxide thicknesses, which were presumably performed in an attempt to improve chemical bonding, were actually strengthening the assembly by realizing lower concentration gradients and hence more favourable stress patterns, similar to a graded seal.

2.4.4 Non-spontaneous metal–glass redox reactions

When an oxide layer is dissolved completely, the oxide saturation at the interface is lost as the diffusion of the dissolved oxide continues into the glass (Fig. 2.7(b)). At this point, because of the loss of chemical equilibrium, redox reactions become possible with the formation of a metal

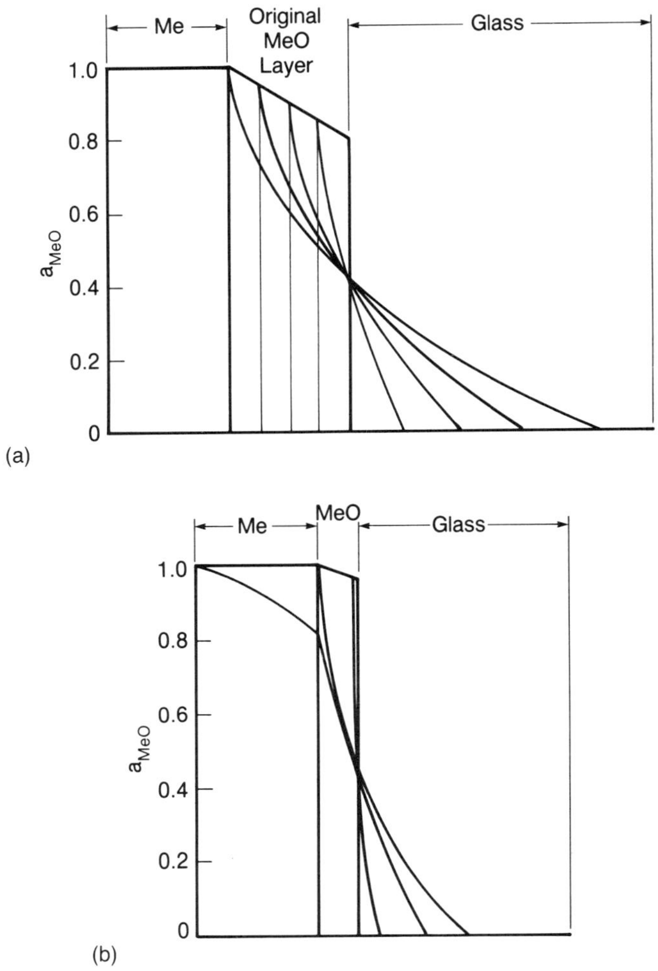

Fig. 2.7 Schematic diagram showing lengthening activity gradients of MeO in glass with continuing solution of oxide by the glass: (a) gradient extension proportional to the thickness of oxide; (b) loss of saturation at the interface with continued heating beyond complete solution of the oxide layer.

oxide at the interface followed by its solution in the glass. For the model system $Na_2Si_2O_5$ glass and iron we can write

$$x\mathrm{Fe} + \mathrm{Na_2Si_2O_5} \rightarrow x\mathrm{FeO_{int}} + \mathrm{Na_{2-2x}Si_2O_{5-x}} + 2x\mathrm{Na}$$
$$\rightarrow \mathrm{Fe}_x\mathrm{Na_{2-2x}Si_2O_5} + 2x\mathrm{Na}\uparrow \qquad (2.4)$$

and the free-energy relationship for the first step can be written

$$\Delta G_{\mathrm{R}} = \Delta G_R^\circ + RT\ln\left[\frac{\{a(\mathrm{FeO_{int}})\}^x\{a(\mathrm{Na})\}^{2x}}{a(\mathrm{Na_2O})}\right]$$

The same basic relationship exists if the glass is applied to an unoxidized metal. The net reaction (eqn (2.7)) consists of two step reactions: the formation of the metal oxide at the interface (eqn (2.5)) followed by its solution by the glass (eqn (2.6)):

$$\mathrm{Fe} + \mathrm{Na_2O_{glass}} \rightarrow \mathrm{FeO_{int}} + 2\ \mathrm{Na}\uparrow \tag{2.5}$$

$$\mathrm{FeO_{int}} \rightarrow \mathrm{FeO_{glass}} \tag{2.6}$$

$$\mathrm{Fe} + \mathrm{Na_2O} \rightarrow \mathrm{FeO_{glass}} + 2\mathrm{Na}\uparrow \tag{2.7}$$

Thus

$$\Delta G_{\mathrm{R}} = \Delta G_R^\circ + RT\ln\left[\frac{a(\mathrm{FeO})_{\mathrm{int}}\{a(\mathrm{Na})\}^2}{a(\mathrm{Fe})\ a(\mathrm{Na_2O})_{\mathrm{glass}}}\right] \tag{2.8}$$

where ΔG° is the free energy of the reaction when the components are in their standard states and a is the chemical activity of the component.

The first step of this reaction does not take place spontaneously at the sealing temperature of 1000 °C (ΔG_{R} is positive) because no cation in this glass has an oxidation potential less than that of the substrate atom. The conditions, other than standard, under which ΔG_{R} for this reaction becomes negative are determined by its equilibrium constant, or activity quotient, shown in equation (2.8). The activity quotient has to be sufficiently smaller than unity for the overall term to be more negative than the positive $\Delta G_{\mathrm{R}}^\circ$. The reaction is thus favoured by a low $a(\mathrm{Na})$, a low $a(\mathrm{FeO})_{\mathrm{int}}$ and a high $a(\mathrm{Na_2O})_{\mathrm{glass}}$. $a(\mathrm{Fe})$ is close to unity and can be neglected. The pressure of the sodium vapour produced must exceed ambient pressure to nucleate bubbles of sodium vapour at the interface and escape; therefore this reaction is potentially more favourable in a dynamic vacuum or a low-pressure ambient atmosphere. The loss of sodium results in a weight loss that can be monitored for analysis of the kinetics of the reaction. The value of $a(\mathrm{Na_2O})_{\mathrm{glass}}$ is dependent on the glass composition, i.e. the availability of Na_2O for reaction. Experimentally, this requirement is fulfilled by $Na_2Si_2O_5$ glass. The $a(\mathrm{FeO})$ at the interface or the iron surface is dependent on the oxygen pressure in the ambient atmosphere and the amount of dissolved FeO in the starting iron. Above the FeO dissociation pressure of 1.2×10^{-15} atm, the surface remains covered with FeO and $a(\mathrm{FeO})_{\mathrm{int}}$ is unity. Below the dissociation pressure, bulk oxide does not form and any oxide that may be present on the surface dissociates to decrease $a(\text{FeO})_{\text{int}}$ below unity. It is necessary to consider the step reactions in order to understand the nature of the net reaction (2.7) fully and to interpret the kinetics correctly. This requirement exists because the second step reaction has a negative $\Delta G^{\circ}_{\text{R}}$ and thus occurs spontaneously.

It is important to be aware of such step reactions because of their possible occurrence as complicating reactions and their effect on processing control factors. In practice, however, it is more convenient to have compositions that lead to spontaneous redox reactions under regular processing conditions, generally in atmospheric air.

2.4.5 Spontaneous metal–glass redox reactions

The porcelain enamel technologists of many generations ago experimentally found that, without changing the firing conditions, a small addition of CoO to the enamel glass caused a spontaneous redox reaction with the substrate metal when all the surface oxide was dissolved. The surface metal oxide was formed either by pre-oxidation or during the initial stages of heating before the applied glass particles flowed:

$$\text{Fe} + \text{CoO}_{\text{glass}} = \text{FeO}_{\text{int}} + \text{Co} \tag{2.9}$$

The change ΔG_{R} for this reaction is negative because the CoO in the glass is easily reduced by iron which has a higher oxidation potential than cobalt (i.e. $Co^{2+} = -0.28\,\text{eV}$ compared with $Fe^{2+} = -0.44\,\text{eV}$). Under normal experimental conditions ΔG_{R} remains negative even though $a(\text{FeO})$ in the iron substrate and interface may be unity. The FeO that is introduced into the glass by either redox reaction ensures that the glass–metal interface remains saturated with the oxide which is necessary to maintain chemical bonding. A contributing factor to a negative $\Delta G^{\circ}_{\text{R}}$ is the energy gain due to the fact that cobalt readily alloys with iron [15,16]:

$$(x + y)\text{Fe} + \text{Co}_y\text{Na}_2\text{Si}_2\text{O}_{5+y} \rightarrow \text{Fe}_y\text{Na}_2\text{Si}_2\text{O}_{5+y} + \text{Fe}_x\text{Co}_y \tag{2.10}$$

This alloying tendency results in the establishment of microgalvanic cells which cause the formation of FeCo_x dendrites in the glass in the interfacial zone, as indicated in Fig. 2.6(b). This reaction, plus the corrosion of the metal interface, leads to the development of a composite-type microstructure which increases the strength of the assembly.

Instead of modifying the glass composition in order to realize a favourable negative $\Delta G^{\circ}_{\text{R}}$ for a redox reaction under any firing conditions, including air at a pressure of 1 atm, it is possible to use a highly reactive metal as a substrate. Experiments have been made with chromium, titanium and zirconium, for example, which have sufficiently high oxidation potentials to reduce even silica in the glass with the formation of

Fig. 2.8 Scanning electron micrograph of the cross-section of the interfacial zone for chromium and glass showing dendritic growth of chromium silicide in the glass at the interface.

an alloy of the metal with silicon or a silicide [17,18]. A representative reaction is that of chromium with $Na_2Si_2O_5$ glass to form platelets and wire-like crystals of $CrSi_x$ alloy (Fig. 2.8) by means of a microgalvanic cell mechanism similar to that for $FeCo_x$ (Fig. 2.6(b)). The released oxygen formed oxides of the substrate metals. Saturation of the interfacial zone with the substrate oxides resulted in chemical bonding. When a titanium substrate was used, a silicide compound (Ti_5Si_3) instead of an alloy ($TiSi_x$) was formed as a layer at the interface. In this case, the compound is compatible with the titanium substrate and the glass.

Another type of glass–metal assembly is one in which the substrate metal is an alloy. Because of the differences in the oxidation potentials or reactivities of the components of the alloys, the redox reactions are not stoichiometric regardless of whether the alloy is exposed to atmospheric oxidation or reacts directly with the glass. Some recent observations on the reactions and bonding of a potassium borosilicate glass to several Ni20Cr alloys (one contained impurities at parts per million levels and a second contained 2.0 wt% Si as the principal impurity) provide examples of the complexities encountered [19]. On pre-oxidation, the purer alloy formed a multilayer oxide scale with an NiO-rich outer layer, whereas the latter alloy formed a chromium oxide scale containing in solution an NiO concentration gradient outward from the alloy interface. The type of cross-

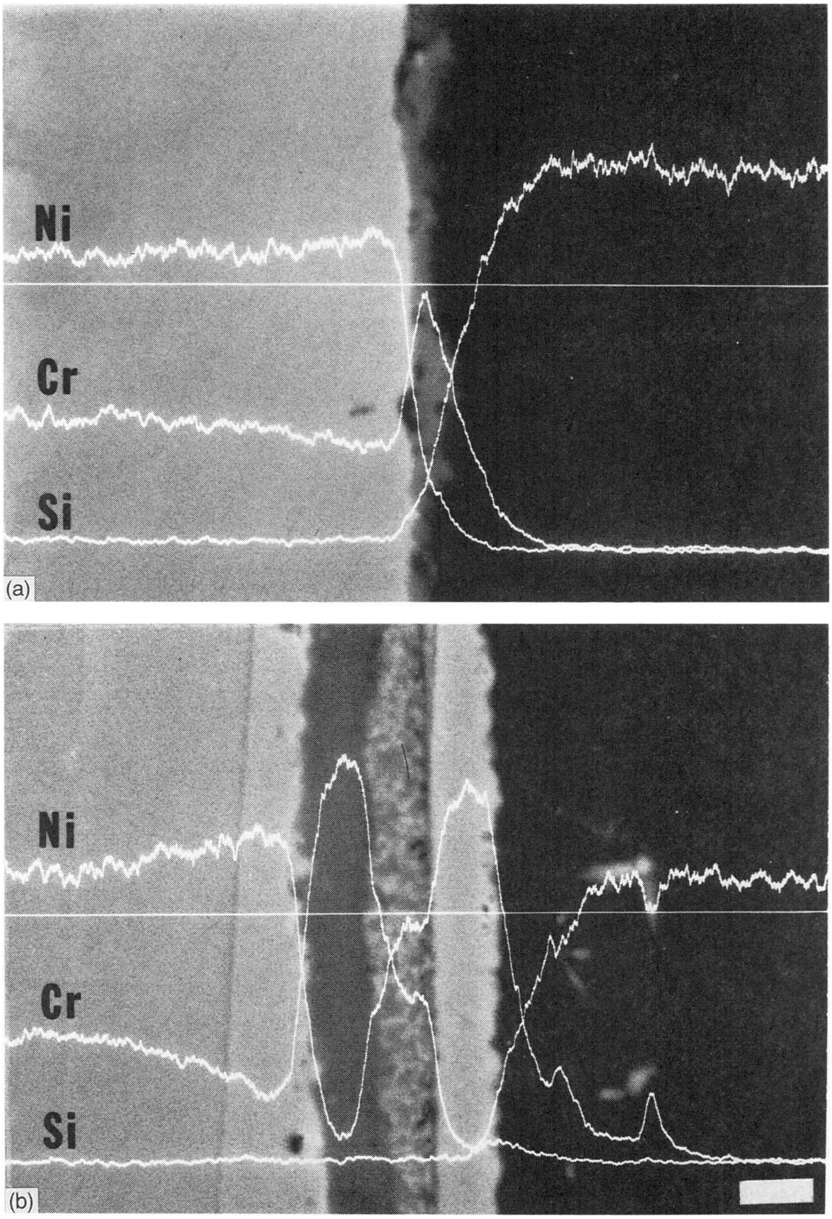

Fig. 2.9 X-ray line scans for chromium, nickel and silicon across the interfaces of the glassy phase of dental porcelain on (a) 80Ni20Cr alloy with silicon additive pre-oxidized for 60 min at 1000 °C in air and (b) 80Ni20Cr alloy with no additive pre-oxidized for 1 min at 1000 °C in air. (Scanning electron micrographs at the same magnification: the bar represents 2 μm.)

sectional oxide microstructures formed can be seen in Fig. 2.9. Stable chemical equilibrium exists at all interfaces. However, bonding and strength problems related to scale adherence integrity arise because of physical incompatibilities such as differences in specific volume, epitaxy and coefficient of thermal expansion.

With application of the glass, solution reactions occur as discussed before. The presence of any undissolved oxide, or any penetration of a porous oxide layer by the glass, plays a significant role in determining the strength and vacuum tightness of the assembly. If all the oxide is dissolved or if the glass is placed in contact with unoxidized alloy, then redox reactions take place with the glass to form predominantly Cr_2O_3 at the interface which is initially dissolved by the glass. With saturation of the glass at the interface, a Cr_2O_3 layer forms, as observed in Fig. 2.9, with the development of chemical bonding. The ultimate acceptability of the assembly will be determined by the physical factors which will be affected by the concentration gradients and microstructures that develop during the chemical reactions that take place in reaching stable chemical equilibrium in the interfacial zone.

2.4.6 Reactivity of metals

In the fabrication of glass–metal seals, glasses come into contact with the metal substrate after dissolving any existing metal oxide on the substrate. Redox reactions are then necessary to maintain the interface saturated with the metal oxide. The reaction at the interface depends on the relative oxidation potentials of the metal and the potentially reducible cations in the glass. The metals can be arbitrarily grouped on the basis of reactivity in a vacuum and in atmospheres with low $a(O_2)$, and hence low oxidation potential, as low, medium and high.

The low-reactivity metals include gold, platinum and silver. Oxidation in air does not produce bulk oxides although adsorbed layers of the oxide of the metal are formed. Such layers do not form in normal vacuum and inert atmospheres, i.e. at low $a(O_2)$, or, if formed, dissociate on heating. When metal–glass assemblies are fired in air, the oxide layer that forms is sufficient to saturate the interfacial zone with the respective substrate oxide because of the very low solubility and diffusivity of these oxides in glasses. In the case of silver, the Ag_2O forms a silver silicate compound at the interface which is compatible with both silver and the Ag_2O–saturated glass. Chemical bonding occurs in all cases. On firing in a vacuum, however, no redox reactions occur at the interfaces of the gold and platinum assemblies, and hence there is no chemical bonding, but redox reactions do occur in the silver–glass assemblies with the development of chemical bonding [20].

The medium group includes such metals as iron, nickel and cobalt. The high-reactivity group includes such metals as chromium, titanium and zirconium. The general chemical behaviour of these types of assemblies has already been discussed.

2.5 THE MAKING OF GLASS–METAL SEALS

Processing factors are important from a technological viewpoint since they require control once proper procedures for making a glass–metal seal have been established. They are also important fundamentally since they can affect the degree of oxidation of the metal surface, the chemical reaction behaviour and the nature of the strain development in the interfacial zone. Thus both thermodynamic and kinetic factors can be affected.

The metals commonly used in making glass–metal seals are platinum, tungsten, molybdenum, copper, iron, nickel and alloys containing various proportions of iron, nickel, cobalt and chromium. Specific sealing glasses are needed for each metal or alloy primarily because of the need for a near match of thermal expansions. Matching combinations of metal and glass can be obtained commercially, and some examples for low-expansion metals – tungsten and Kovar – are given in Table 2.1.

When selecting a glass to match a metal workpiece, it is prudent to discuss the particular application with the supplier since the optimum selection will depend on a number of specific properties and requirements. In general, the coefficients of thermal expansion should be within 10% of each other, with that of the metal being the higher. The contraction of interest is that from the setting temperature of the glass to room temperature. If the difference in contraction is less than 100 ppm, the stress level will be quite safe, and even 500 ppm is usually satisfactory. A seal with a contraction difference between 500 and 1000 ppm can still be used but only for progressively smaller and thinner assemblies. However, a

Table 2.1

Glass supplier	*Identification number*
Tungsten	
Corning (USA)	7720,3320,7750
Glass Bulbs (UK)	B37
Schott (FRG)	8486, 8487
Kovar	
Corning (USA)	7052
Glass Bulbs (UK)	B47
Schott (FRG)	8454, 8436

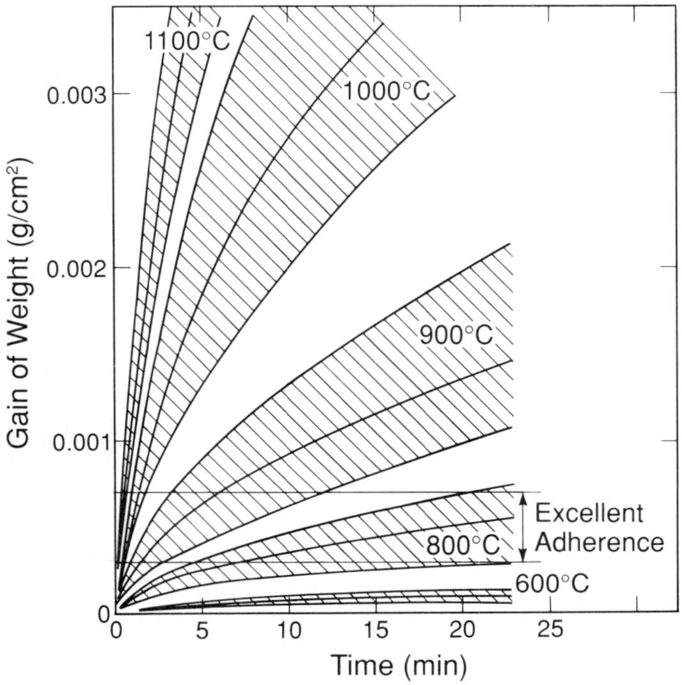

Fig. 2.10 Oxidation rate of small pieces of Kovar as a function of time and temperature.

close match in expansion coefficients is not as critical for thin glass coatings as for bulk specimens.

In preparation for a glass–metal seal, the metal surface is thoroughly cleaned of all material that may be insoluble or may react with the glass at the sealing temperature (primarily carbon-containing material) because such material normally interferes with the wetting of the metal by the glass. In addition, carbon forms CO and CO_2 on reaction with oxides in the glass. These reactions produce bubbly seals that do not have a very good appearance, are weak and may not be hermetic if the bubbling is severe. The usual method of removing carbon is to oxidize it by heating the metal in a wet hydrogen atmosphere to about 1100 °C. The cleaning process may also result in some etching or roughening of the metal surface. The clean metal is then pre-oxidized if the amount of oxidation that will occur during the sealing process is insufficient. Figure 2.10 describes the oxidation of Kovar, a Co–Ni–Fe alloy that is used extensively in glass–metal seals [8].

After cleaning, the bulk glass is heated and applied to the heated metal. Glass is heated with the metal when it is applied as a powder to form a protective coating or as a thin glass primer to which the bulk glass is subsequently sealed, or when it is a pressed powder glass preform in a metal frame with pins inside the preform. In all cases, it is essential that at the sealing temperature the glass should dissolve most, if not all, of the surface oxide on the metal to form a bond with the metal. As shown in Fig. 2.10, the amount of oxide, and not the time and temperature of its formation, is the principal factor in developing good adherence. If the oxide layer is thin, the seal is weak (although it may be vacuum tight, especially under compressive stresses). If the oxide layer is thick, some oxide may remain after sealing, and the strength of the seal is thus affected by the strength of the oxide and its adherence to the metal. Also, if the oxide layer is continuous and porous, the seal may not be vacuum tight. It should be noted that the colour of the glass at the interface at room temperature provides some indication of the quality of the seal. For example, in good Kovar–Corning 7052 glass seals with the proper amount of oxide, the colour at the interface is dark bluish grey. This colour is unaltered if the oxide layer is too thick, but becomes lighter if there is insufficient oxide.

After the seal has been fabricated, the assembly is cooled to room temperature. The critical range for cooling is just below the transformation temperature range of the glass. The assembly must be cooled slowly, or annealed, through this range in order to avoid destructive thermal strains in the glass. The importance of annealing increases with increasing thickness of the glass.

2.6 CONCLUSIONS

In this chapter we have dealt primarily with the science and technology of glass–metal seals. Both physical and chemical factors are involved in making seals. The aim of the discussion has been to develop an understanding of the functions of these factors in producing seal assemblies that are strong and vacuum tight, fulfil their function, and are reliable and reproducible.

The materials are selected on the basis of having desirable properties and characteristics for the application. The same principles and requirements apply to the making of every seal. Engineering and physical characteristics are of concern because correct selection can ensure acceptably low stress gradients in the interfacial zone. Similarly, chemical characteristics are of concern because of their role in producing a strong bond at the interface and a favourable microstructure in the interfacial zone.

The stress gradients and a chemical bond are critical because they determine the strength of the overall assembly. The stress gradients develop on cooling and are controlled by the difference between the coefficients of thermal expansion of the glass and the metal. A chemical bond across the interface results in the strongest bond. It is achieved by controlling chemical reactions at the interface so that thermodynamically stable equilibrium compositions are obtained at the interface. With oxide glasses this corresponds to saturation of the interfacial zone with the lowest-valent oxide of the substrate metal which will be compatible with both the phases being bonded. The extent of the reaction determines the width of the reaction zone and hence the stress gradients.

It is evident that the physical and chemical factors do not operate independently in influencing the technology of glass–metal seals. Furthermore, the processing procedures have to be adjusted in order to realize the best conditions for obtaining the optimum seal. The strength of the overall assembly and its integrity and durability in an application are the dominant factors in their development and selection.

REFERENCES

1. Housekeeper, W.G. (1923) The art of sealing base metals through glass, *J. Am. Inst. Electr. Eng.*, **42**, 870–6.
2. Schott Product Information, No. 4830e, Glass-to-Metal Seals, Schott, Jena Glasswerk Schott and Gen. Inc., New York.
3. Borom, M.P. and Johnson, C.A. (1989) Thermomechanical mismatch in ceramic-fiber-reinforced glass–ceramic composites, *J. Am. Ceram. Soc.*, in press.
4. Bertolotti, R.L. (1983) Porcelain to metal bonding and compatibility, in *Dental Ceramics, Proc. 1st Int. Symp. on Ceramics*, (ed. J.W. McLean), Quintessence, San Francisco, pp. 415–31.
5. Humenik, M., Jr, and Whalen, T.J. (1960) Physiochemical aspects of cermets, in *Cermets* (eds J.R. Tinklepauch and B.W. Crandall), Reinhold, New York, pp. 6–49.
6. Tomsia, A.P. and Pask, J.A. (1986) Chemical reactions and adherence at glass/metal interfaces: an analysis, *Dent. Mater.*, **1** (2), 10–16.
7. Klomp, J.T. (1987) Ceramic–metal reactions and their effect on the interface microstructure, in *Ceramic Microstructures '86: Role of Interfaces* (eds J.A. Pask and A.G. Evans) Plenum, New York, pp. 307–17.
8. Pask, J.A. (1948) New techniques in glass-to-metal sealing, *Proc. IRE*, **26** (2), 286–9.
9. Aksay, I.A., Hoge, C.E. and Pask, J.A. (1974) Wetting under chemical equilibrium conditions, *J. Phys. Chem.*, **12** (78) 1178–83.
10. Sharps, P.R. Tomsia, A.P. and Pask, J.A. (1981) Wetting and spreading in the Cu–Ag system, *Acta Metall.*, **29** (7), 855–65.
11. Tomsia, A.P. and Pask, J.A. (1981) Kinetics of iron sodium disilicate reactions and wetting, *J. Am. Ceram. Soc.*, **64**(9), 523–8.
12. Brennan, J.J. and Pask, J.A. (1973) Effect of composition on glass–metal interface reactions and adherence, *J. Am. Ceram. Soc.*, **56** (2), 58–62.
13. Hoge, C.E. Brennan, J.J., and Pask, J.A. (1973) Interfacial reactions and wetting behavior of glass iron systems, *J. Am. Ceram. Soc.*, **56** (2) 51–4.
14. Mayer, P., Topping, J.A. and Murthy, M.K. (1974) Correlation between

thermal expansion of glass and glass to metal adherence, *J. Can. Ceram. Soc.*, **43**, 43–6.
15. Borom, M.P. and Pask, J.A. (1966) Role of adherence oxides' in development of chemical bonding at glass–metal interfaces, *J. Am. Ceram. Soc.*, **49** (1), 1–6.
16. Borom, M.P., Longwell, J.A., and Pask, J.A. (1967) Reactions between metallic iron and cobalt oxide bearing sodium disilicate glass, *J. Am. Ceram, Soc.*, **2**(50), 61–6.
17. Pask, J.A. and Tomsia, A.P. (1981) Wetting, spreading and reactions at solid/liquid interfaces in *Surfaces and Interfaces in Ceramic and Ceramic–Metal Systems* (eds J.A. Pask and A.G. Evans) pp. 411–19, Plenum, New York.
18. Tomsia, A.P., Zhang, F. and Pask, J.A. (1985) Reactions and bonding of sodium disilicate glass with chromium, *J. Am. Ceram. Soc.*, **68** (1), 20–4.
19. Tomsia, A.P. and Pask, J.A. (1986) Bonding of dental glass to Ni–Cr alloys, *J. Am. Ceram. Soc.*, **69**(10), C239–40.
20. Nagesh V.K., Tomsia, A.P. and Pask, J.A. (1983) Wetting and reactions in the lead borosilicate glass–precious metal systems, *J. Mater. Sci.*, **18** (11), 2665–70.

3

Joining glass-ceramics to metals

G. Partridge

3.1 INTRODUCTION

The advancement of engineering technology to meet future requirements in many diverse fields demands the development of new ceramic materials. No single ceramic material can meet all needs and consequently many different types of material require investigation involving the tailoring of composition and processing to meet specific requirements. An important class of materials within the overall range of ceramic materials is that of glass-ceramics which are made by the controlled crystallization of inorganic glass compositions.

For many years glass workers aimed at developing compositions which would not crystallize, or devitrify, during shaping or processing operations or in use, and the many commercially available glasses are witness to their efforts. The incidence of crystalline phases, which were present as 'gross' crystals, meant a deterioration of the workability of the glass and in its properties, particularly strength and transmission. It was appreciated, however, that if the crystallization of a glass could be controlled then useful products might result. Indeed, as long ago as 1739, Réaumur [1] attempted to produce a 'porcelain' material by heat-treating soda-lime–silica glass bottles immersed in a mixture of sand and gypsum. Although crystallization was achieved, the process was uncontrolled and the crystallization was the result of devitrification from the surface which resulted in weak and brittle materials.

In the 1950s further research led to the realization that, under suitable conditions and, at that time, in a restricted range of compositions, it was possible to produce a usable polycrystalline material by the controlled devitrification of a glass provided that sufficient nuclei on which the crystals could grow were developed in the glass [2]. The means whereby this can be achieved and examples of typical composition ranges are discussed in the next section.

In common with ceramics made by sintering processes, it is often necessary for glass-ceramics to be joined to other components, such as metals, in order to provide required devices or parts of devices. The present chapter is concerned with the preparation of bonds between glass-ceramics and metals, and a number of working examples are reviewed as an aid to understanding the principles involved in the selection of materials and fabrication of the joins.

3.2 THE GLASS-CERAMIC PROCESS

The preparation of a glass-ceramic involves a number of steps. The first of these, which gives rise to the generic name for these materials, is the preparation of a glass. The mixture of raw materials which will form the required composition is melted in a suitable container at a temperature generally in the range 1000–1700 °C (depending upon composition). After refining and homogenization to give good quality glass, the molten glass is worked by the usual methods employed in the glass industry, e.g. casting, pressing, drawing, or chilling to form glass frit which can subsequently be reduced to powder. The majority, although not all, of the compositions which form glass-ceramics are silicate based, and the ranges of compositions which can be melted and which can then be converted to the polycrystalline structure of a glass-ceramic are now extensive [3,4]. The preparation of a satisfactory glass-ceramic from the bulk glass depends on crystallizing the glass composition under strictly controlled conditions in order to provide the required closely interlocking microcrystalline structure and a smooth uncrazed surface. In order to do this it has generally been necessary to include a material in the glass which will provide nuclei for subsequent crystal growth, or to control the crystal formation in such a manner that many crystals of the desired type(s) grow in the glass. Oxide nucleants such as P_2O_5, TiO_2 and ZrO_2 are preferred in many instances, although in some cases a combination of two or more nucleants has been shown to give better results than any one individual nucleant used in the same total proportion. Normally, the glasses in bulk form are subjected to a controlled heat-treatment cycle (as shown in Fig. 3.1) which converts the glass to glass-ceramic. The heat treatment comprises nucleation and crystallization temperature holds, and more holding stages may be included as necessary to develop the desired structure and properties. The heat-treatment process is designed such that the microstructure of the resultant material is one in which one or more crystal phases exist together with a residual glassy phase. A closely interlocking microstructure results in mean crystal sizes generally in the region of 1 μm [3,4], although in some cases the mean crystal size can be considerably less [5].

Alternatively, the glass can be reduced to powder and suitably compacted by, for example, die or isostatic pressing and then sintered, essentially in a similar manner as for a conventionally sintered ceramic (e.g. alumina), to provide a preform for use in subsequent processing. The

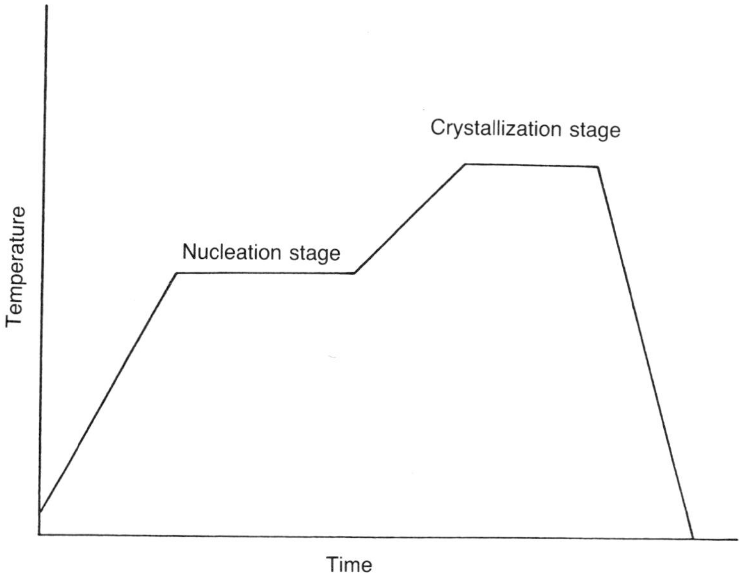

Fig. 3.1 Typical heat-treatment schedule for the production of glass-ceramics showing nucleation and crystallization holding stages.

powder particles are generally in the range 1–30 μm, although this can vary depending on the degree of comminution achieved. The surfaces of the particles are in a highly damaged state and consequently the glass particles are highly susceptible to surface crystallization. However, if extensive surface crystallization occurs, the sintering of the powder to form a vitreous body may be inhibited [6,7] leading to an inferior product. Two possibilities exist in the use of sintered glass materials. In the first, the compact is sintered to a vitreous body and is subsequently heat treated as in the case of bulk materials discussed above. Alternatively, nucleation and bulk crystallization occur during the same firing cycle as is used for the sintering step; this is to be preferred from an economic point of view.

The subsequent glass-ceramics made either by crystallization of bulk materials or by the sintering of powder materials possess totally different properties to those of the original glasses. Mechanical properties are often improved, particularly for bulk heat-treated materials, and refractoriness is increased by several hundred degrees Celsius compared with the parent glass. In addition, good electrical and thermal properties ensue. It is not within the scope of the present chapter to discuss the properties of glass-ceramics in detail except in so far as they are relevant to the development of specific glassceramic–metal bonded composite structures. Details of typical glassceramic properties can be found in a number of publications [3,4] and from manufacturers' data sheets.

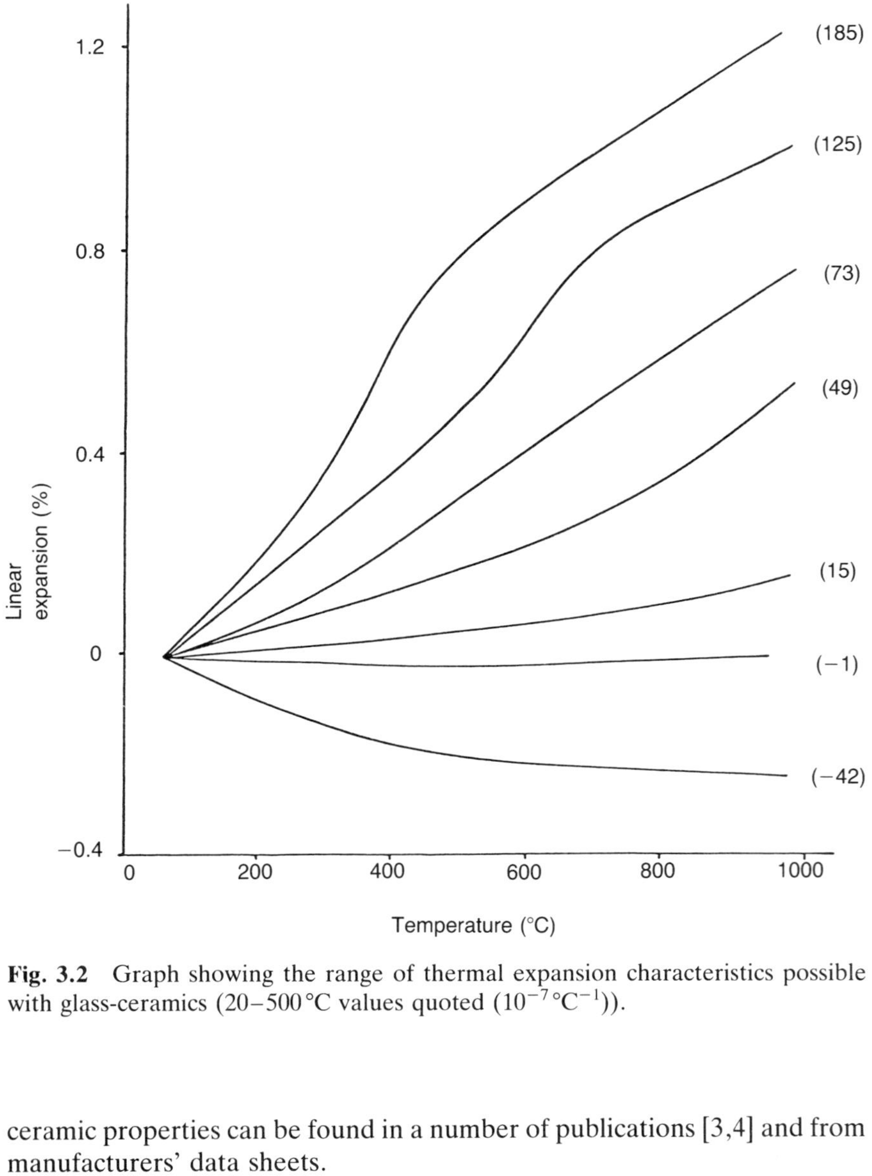

Fig. 3.2 Graph showing the range of thermal expansion characteristics possible with glass-ceramics (20–500 °C values quoted (10^{-7} °C^{-1})).

ceramic properties can be found in a number of publications [3,4] and from manufacturers' data sheets.

An important feature of glass-ceramics is that, out of the extensive range of compositions which can be prepared, it is possible to produce materials covering a wide range of thermal expansion characteristics as illustrated in Fig. 3.2. These range from negative through zero to highly positive expansions similar to those of, for example, copper. From this range it is possible to select materials which are matched in expansion to many metals and alloys. Examples of thermal expansion matching of glass-ceramics to

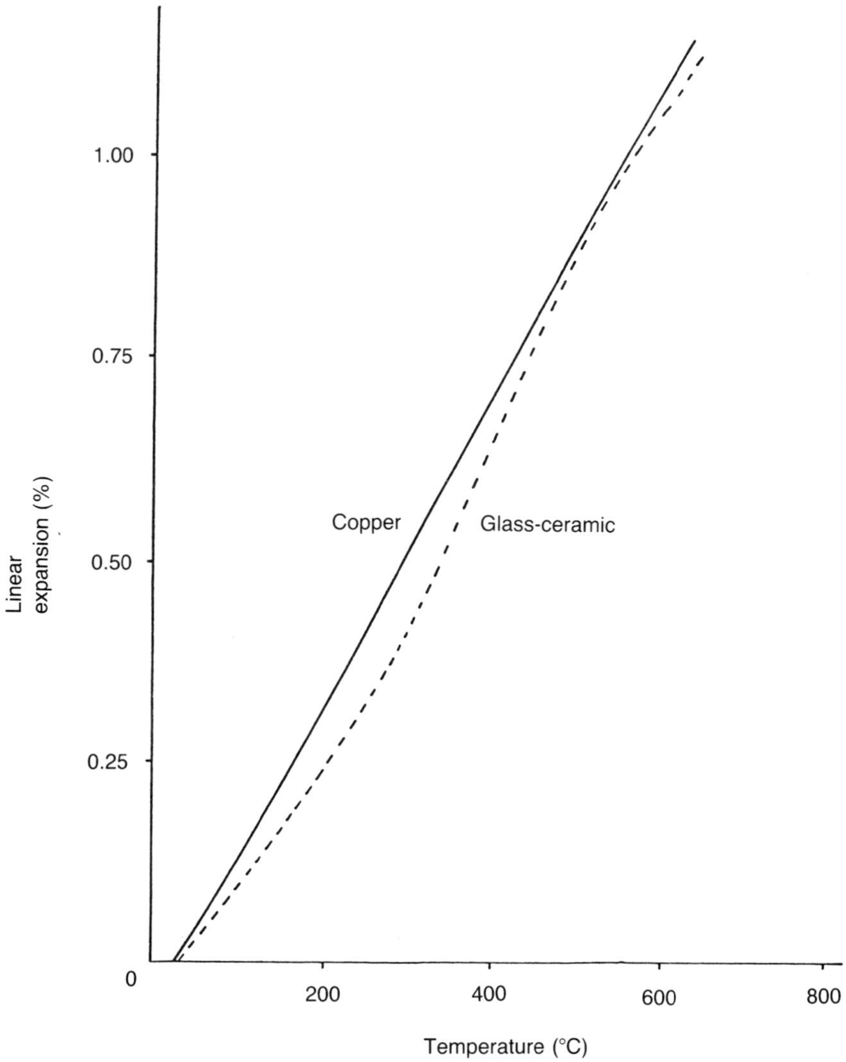

Fig. 3.3 Thermal expansion characteristics of copper and matching glass-ceramic.

various metals and alloys are shown in Figs. 3.3–3.5, and the importance of thermal expansion matching is discussed further in subsequent sections.

3.3 BONDING GLASS-CERAMICS TO METALS

There are a number of ways by which a high quality bond between a glass-ceramic and a metal or alloy can be achieved. These are employed when:

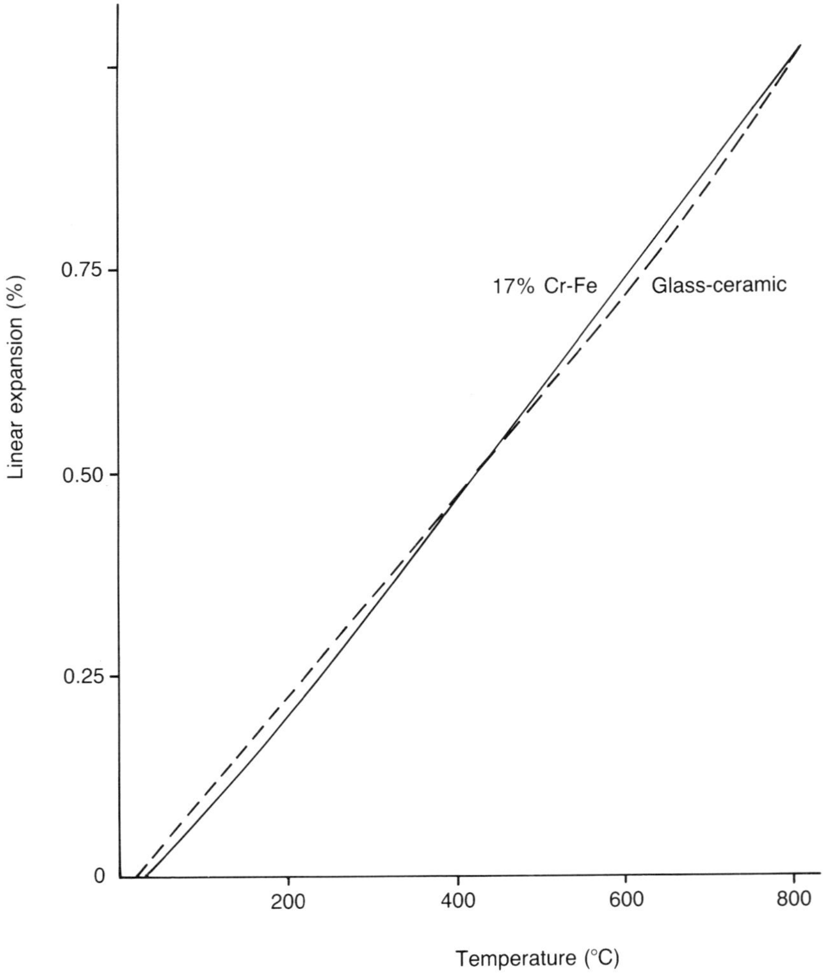

Fig. 3.4 Thermal expansion characteristics of 17%Cr–Fe and matching glass-ceramic.

1. joining bulk components
   (a) using an intermediate bonding medium
   (b) direct bonding;
2. coatings.

In many instances, particularly where joining of bulk components is required, good thermal expansion matching of the components is necessary in order to ensure low stress levels in the bond and maintain bond integrity at temperatures which may be experienced during manufacture, processing

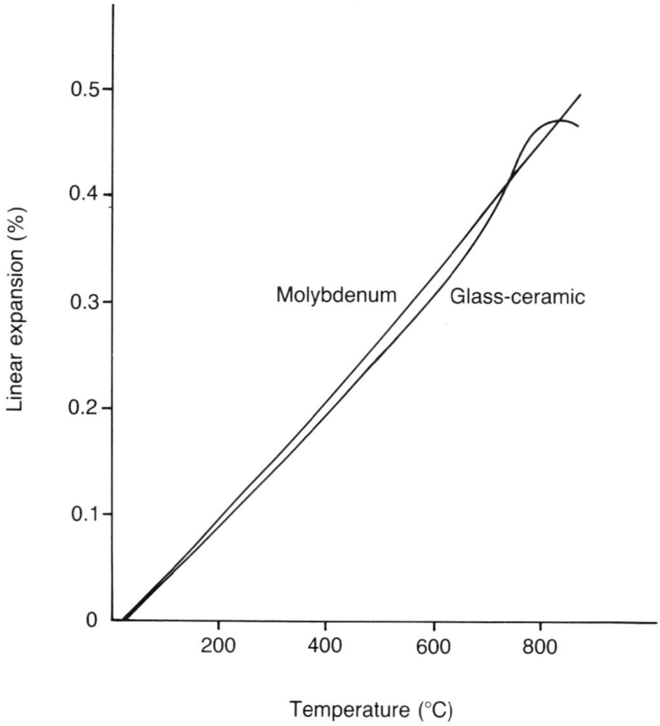

Fig. 3.5 Thermal expansion characteristics of molybdenum and matching glass-ceramic.

and in use. In some cases, however, compressive stresses are deliberately introduced into the glass-ceramic component in order to enhance its mechanical properties, and this is particularly relevant to certain coating applications. Of course, whilst the thermal expansion matching situation is of paramount importance and has a considerable influence on the choice of glass-ceramic for a given combination and application, it is important to appreciate that other factors must be taken into account. Firstly, the chosen glass-ceramic must be capable of being bonded to the proposed metal or alloy. In the case of direct bonding this further implies that, in its initial glassy form, the glass-ceramic must wet and adhere to the metallic element. Additionally, the other properties of the glass-ceramic, e.g. mechanical strength, electrical resistivity and dielectric properties, must meet the overall specification for the product.

The techniques employed for the preparation of the various bonds between glass-ceramics and metals are given below, together with examples of such bonded components.

3.3.1 Joining using an intermediate bond

Two types of intermediate bond can be employed, and the selection of which to use will be influenced by a number of practical features including the ease of fabrication, the temperature-withstand capabilities required, whether or not the bond must possess insulating properties and the cost.

The bonds can be either metallic (e.g. a braze) or vitreous (glass or glass-ceramic). In this type of bond the usual design configuration is that of a butt seal in which the glass-ceramic components are joined on their end-faces to the metallic components, although other types of seal, such as sleeve or concentric seals, are possible.

The glass-ceramic body components are initially prepared using glass-forming methods and are then heat treated. The butt type of seal usually requires that the end-faces of the glass-ceramic components are machined flat and parallel, and other surfaces, which are very often inner and outer cylinder surfaces, are machined to tight tolerances to 'fit' the metal parts. The metal parts themselves also require preparation before assembly. In butt seals the sealing faces must be flat to ensure good mating with the glass-ceramic seal faces. The preparation given to the metal prior to sealing depends on the type of intermediate bond to be employed.

(a) Bonding by means of a metallic braze

In the case of a braze metal bond the metal seal parts are required to have smooth sealing surfaces and to be clean and free from oxidation and other defects. The type of braze normally employed with a glass-ceramic is an

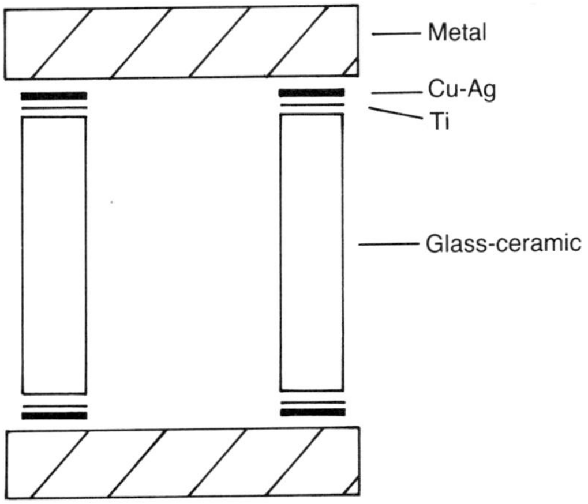

Fig. 3.6 'Exploded' view of the constituent parts for an active metal brazed glass-ceramic–metal seal.

active metal braze. In the butt seal design, thin washers of titanium and Cu–Ag eutectic are assembled together with the major components as illustrated in Fig. 3.6. The titanium is placed adjacent to the glass-ceramic and the Cu–Ag eutectic adjacent to the metal, the whole being in suitable jigs to provide and maintain alignment. Alternatively, a specially developed Cu–Ag–Ti braze alloy can be used [8]. The firing is carried out in a vacuum in order to prevent any oxidation of metallic constituents which would inhibit wetting and flow of the braze at a temperature of about 850 °C. The titanium serves to react with the glass-ceramic during this firing, metallizing the surface and permitting reaction with the braze, which in turn wets and bonds to the metal surface. This type of bond permits subsequent processing (e.g. vacuum bake-out) and operation to temperatures in the region of 800 °C. Higher temperature-withstand capability can be achieved by using more refractory brazes such as those containing nickel.

However, it will be appreciated that the limitation may become the glass-ceramic component itself, although glass-ceramics are available which will withstand temperatures well in excess of 1000 °C. It should be noted that the more refractory glass-ceramics generally have relatively low thermal expansion and consequently matched expansion systems will only be possible with the lower-expansion metals and alloys, e.g. molybdenum and tungsten. It should be borne in mind that work at elevated temperatures may require the presence of an inert atmosphere to prevent oxidation and deterioration of metal parts, including the braze.

(b) Bonding by means of a vitreous bond

In the case of a vitreous bond (glass or glass-ceramic), the sealing faces of the metallic component are roughened by means of grit-blasting or etching and the components are pre-oxidized by heating in a suitable atmosphere. The metal sealing faces are then coated with the vitreous bond which is applied in powder form via a suitable suspension, usually by spraying. The coating is allowed to dry and is then fired, often in a protective atmosphere to protect the metal from excessive oxidation, to form a continuous adherent layer on the metal sealing face. It is usually unnecessary to preglaze the glass-ceramic sealing faces, although this can be done, in a similar manner, if it is considered desirable.

The various components are assembled as shown in Fig. 3.7 for a butt seal and are then fired to cause the bonding materials to complete reactions with the glass-ceramic and metal components and so form the desired bond. If a glass-ceramic bond is employed, the firing schedule will also cause the glass to crystallize. A glass bond will permit subsequent processing and operation to temperatures up to about 500 °C, depending on the glass used, and a glass-ceramic bond will enable temperatures in

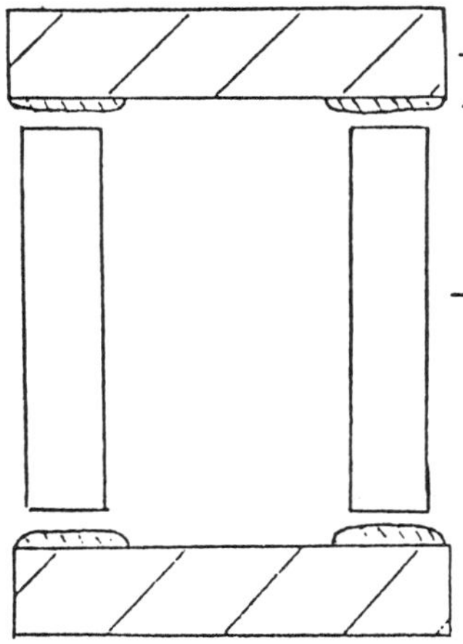

Fig. 3.7 'Exploded' view of the constituent parts for glass- or glass-ceramic-bonded glass-ceramic–metal seal.

excess of 800 °C to be withstood, again depending on the glass-ceramic used.

It is important to note that in using very thin bonds of this type the major consideration has to be given to the thermal expansion match between the main glass-ceramic and metal components, and the vitreous bond needs to be considered more in terms of the 'setting characteristics' that it will impart to the joint and not so much to its expansion matching. On cooling from the bonding temperature, a 'setting temperature' for the joint itself will be reached which can be defined as that below which the bonding glass or glass-ceramic is of so high a viscosity that effective stress relief cannot take place in a reasonable time. This temperature can usually be equated to the annealing temperature applied to the assembly to ensure the desired low-stress relationship between the major components.

An example which illustrates the need to consider the thermal expansion matching of the bulk components in the assembly rather than the bond is the glass-ceramic to mild steel butt seal made to the design shown in Fig. 3.8. The various components have the thermal expansion characteristics shown. In this case the major components are more closely matched with the use of glass B than with glass A, although glass B does not appear so well matched in the system as glass A. Seals made using glass B, however, are significantly stronger than those made using glass A [9]. This works well in this particular case, but it should be appreciated that, if the mismatch between the bond material and the glass-ceramic or metal is too great, then failure may occur owing to the high shear stresses generated at the interfaces.

The use of vitreous bonds is advantageous where more than two

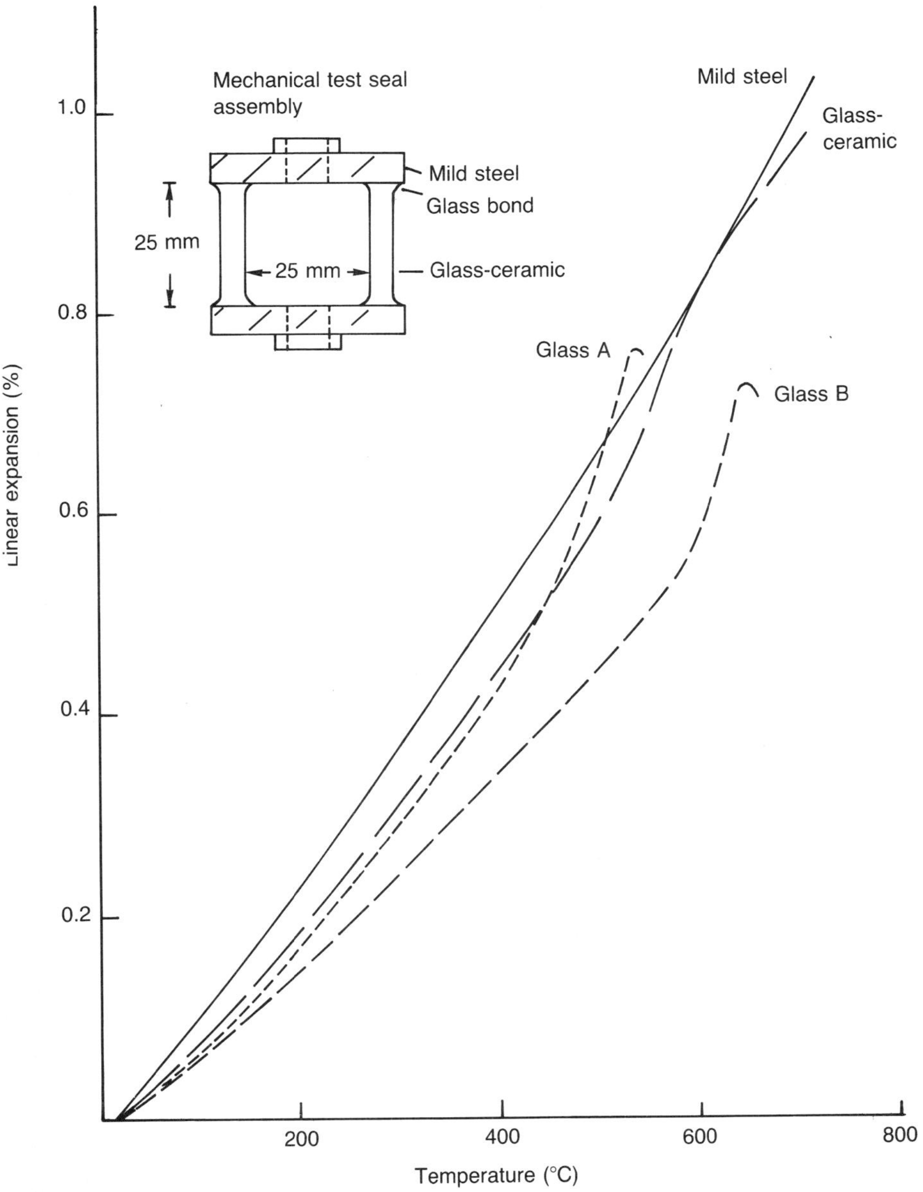

Fig. 3.8 The thermal expansion characteristics of mild steel, glass-ceramic and bonding glasses A and B.

components are required to be bonded together in such a manner that sequential bonding operations are required. It is necessary in these circumstances, while maintaining thermal expansion matching/setting temperature criteria, that the second and subsequent glasses are used at successively lower temperatures so that the firing conditions do not 'disturb' the preceding glass and thereby allow relative movement of the previously bonded components. Fortunately, glasses are available which permit this sequential operation to be achieved, particularly for higher thermal expansion systems. If necessary, a glass-ceramic or braze can be used to provide the initial, more refractory bond.

3.3.2 Direct bonding

As noted above, advantage can be taken of the fact that glass-ceramic materials are initially glasses and as such can be caused to react directly with and bond to suitable metal surfaces. In all cases, the usual design configuration is that of a sleeve or concentric seal in which the glass-ceramic components are joined on cylindrical faces to the metallic components. A major advantage of this type of seal is that the glass-ceramic components do not require precision machining to size and/or flatness (on sealing faces) as in the butt seals described above. Overall, machining is kept to a minimum. As with the vitreous intermediate bonds discussed above, the sealing faces of the metal components are usually roughened by grit-blasting or etching and are then often pre-oxidized by heating in a suitable atmosphere.

Two different methods have been developed to prepare bulk glass-ceramic to metal joins of this general type.

(a) In-mould shaping

For in-mould shaping the sequence of operations is essentially as shown in Fig. 3.9 [10]. The metal parts are positioned in the preheated glass-shaping mould. Molten glass is introduced into the mould and formed to shape. A pressing method is shown in the figure, but a technique such as centrifugal casting could equally well be employed to shape the molten glass. When the glass–metal assembly so formed is sufficiently cool that it will not deform on removal from the mould, it is transferred to a furnace where it is subjected to a heat-treatment process using the type of schedule described earlier. This converts the glass to glass-ceramic, and a glass-ceramic–metal seal assembly is now available. Minor machining operations may have to be carried out on the glass-ceramic, for example to remove a residual glass web in a pressed base and so provide a cylindrical assembly.

In this method, because the time–temperature cycle is limited when the molten glass contacts the metal in the glass-shaping mould, it may be

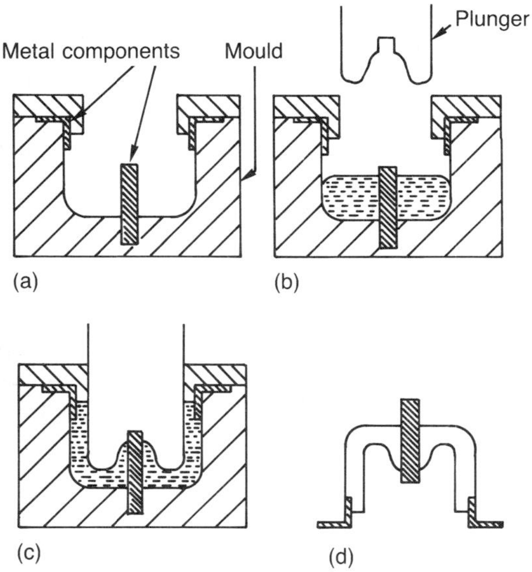

Fig. 3.9 Stages in the preparation of glass-ceramic–metal seals using molten glass.

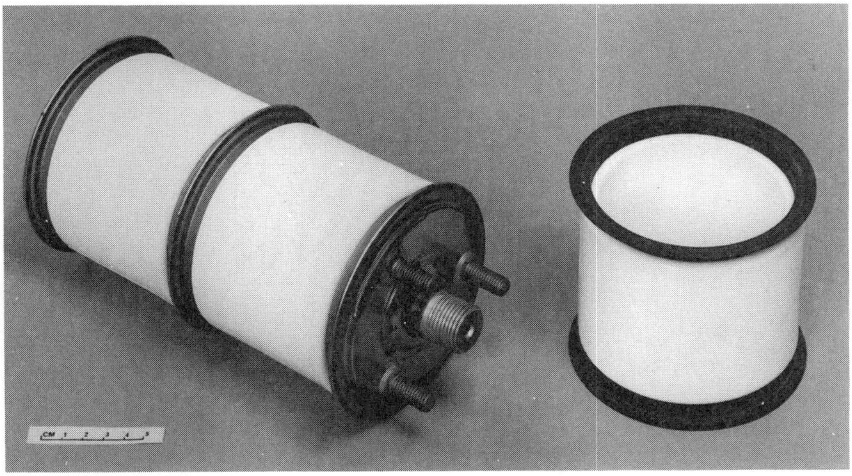

Fig. 3.10 Glass-ceramic bonded to 17%Cr-Fe to provide envelope sections for vacuum interrupters.

advantageous to precoat the sealing faces of the metal components. The coating used is generally very similar to the composition of the bulk material, and it is applied and fired on in a manner analogous to the application of a vitreous bond as described previously.

An example of a glass-ceramic–metal seal assembly made by this technique is shown in Fig. 3.10. This comprises a glass-ceramic of the lithium aluminosilicate type matched in expansion to and bonded to 17%Cr–Fe pressings (a relatively cheap, readily available and corrosion-resistant alloy). The linear thermal expansion match of the major constituents is illustrated in Fig. 3.4. The glass-ceramic also exhibits high strength (modulus of rupture, about 350 MPa) and good electrical characteristics. These seals are manufactured for use in the production of devices where an envelope of high vacuum and mechanical integrity is required.

(b) Graphite jig shaping

In graphite jig shaping the sequence of operations is essentially as shown in Fig. 3.11 [10]. Glass parts are prepared by the shaping of bulk glass to the desired form, for example by pressing from the molten glass and machining as necessary to provide a 'loose' fit with the metal parts and/or the jig. Alternatively, the glass can be reduced to powder, pressed to shape using an appropriate binder and then sintered to provide a component which can be handled.

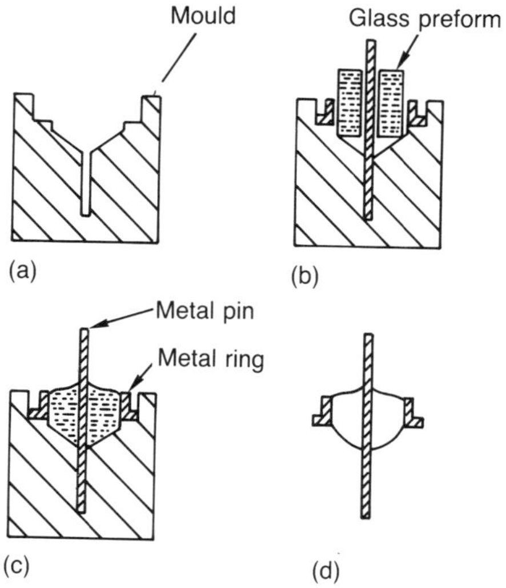

Fig. 3.11 Preparation of glass-ceramic–metal seals using a graphite jig glass reshaping technique.

Metal parts are pretreated in a similar manner to that mentioned above to provide an adherent oxide layer on sealing faces.

The glass and metal components are assembled, often in a suitable graphite jig (as in Fig. 3.11). The assembly is then heated in a furnace to a temperature at which the glass softens sufficiently to flow into contact with the metal parts. The sealing temperature is maintained for a sufficient time to ensure that the desired bond is developed and the assembly is allowed to cool to room temperature. Firings are usually carried out in a protective atmosphere (e.g. nitrogen or argon) to protect metal parts and the graphite jigs from undue oxidation leading to a weakening of the bond. For the majority of the glass-ceramics used in this type of direct bond, nucleation and crystallization occurs in the same step as that used to effect the bond between the glass-ceramic and the metal so that heat treatment as discussed previously is unnecessary. It may be necessary to introduce a hold at a suitable temperature during the cooling cycle to complete crystallization in the material. However, it is important that the crystallization is not too rapid and does not occur too readily during the heating cycle as this will hinder the development of the desired bond. The presence of excessive bulk crystallization will effectively increase the viscosity of the material [11] and impede flow unless higher temperatures are used. The presence of surface crystallization in a powder body will inhibit sintering [6,7] and will also inhibit the reaction of the glass preform with the metal. In many cases the flow and sealing process can be assisted by application of a suitable load onto the softening glass to force the glass into better contact with the metal. This loading can be produced by means of weights or by the use of centrifugal forces.

Examples of glass-ceramic–metal joints produced using this route are as follows.

1. The glass-ceramic–molybdenum high-voltage feedthrough seal illustrated in Fig. 3.12 comprises a glass-ceramic of the $ZnO-Al_2O_3-SiO_2$ type which is matched in expansion and bonded to the molybdenum parts to form the concentric design shown [9]. The linear thermal expansion match of the major constituents is shown in Fig. 3.5.
2. Glass-ceramic–metal leadthrough seals (e.g. in thermocouple terminations) are illustrated in Fig. 3.13. In this type of seal assembly the glass-ceramic is matched in expansion to the metal(s) of the leadthroughs. In a thermocouple termination these would be Chromel and Alumel. In other types of light current leadthrough seals the leadthrough metal is often an Ni–Fe alloy, e.g. 48%Ni–Fe. In this type of seal a higher-expansion metal or alloy, e.g. mild steel or a stainless steel, is frequently used for the outer metal component. A fully hermetic bond is developed between the glass-ceramic and the outer metal and, owing to the differential expansion, or more precisely contraction, of the two

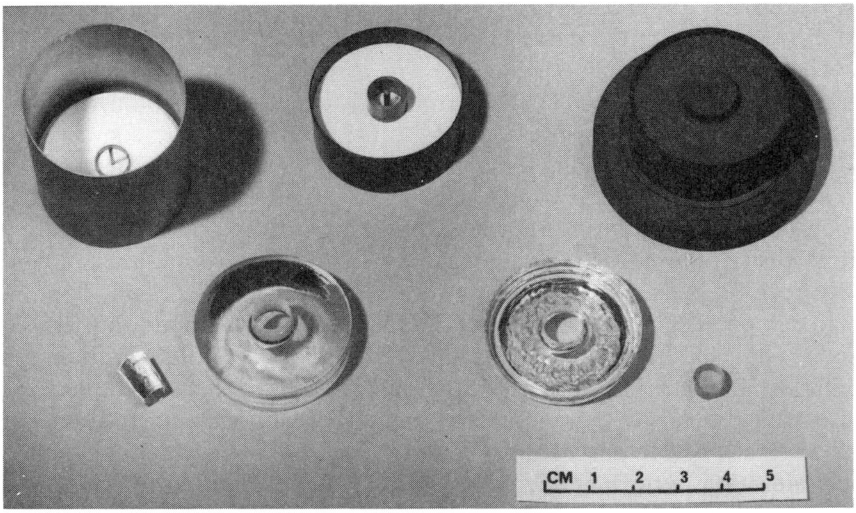

Fig. 3.12 Glass-ceramic–molybdenum high-voltage feedthrough seal. (Courtesy of the Atomic Weapons Research Establishment, Aldermaston.)

components, the outer join is in compression which increases the overall strength of the assembly and its ability to resist thermal shock. Clearly, the deliberate introduction of compressive stresses into the outer glass-ceramic influences the design of the seal assembly. The design must not be such that tensile stresses could be generated in particular directions which would result in failure of the seal assembly,

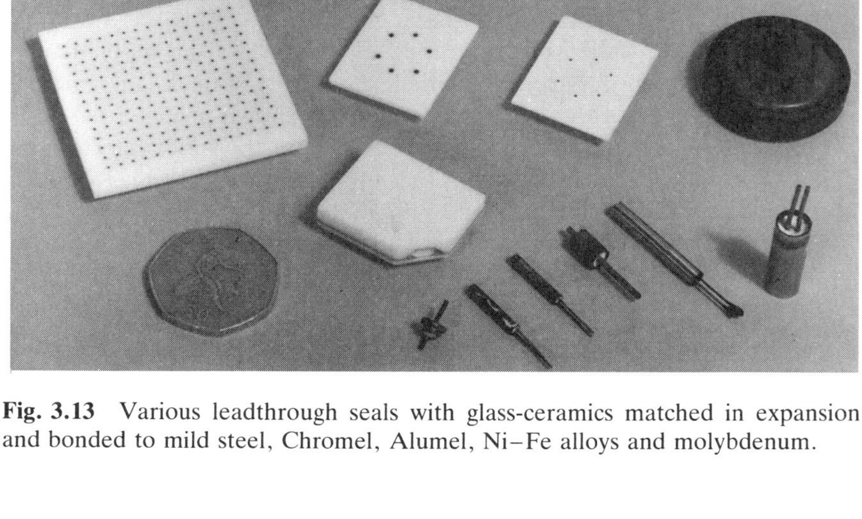

Fig. 3.13 Various leadthrough seals with glass-ceramics matched in expansion and bonded to mild steel, Chromel, Alumel, Ni–Fe alloys and molybdenum.

usually seen as a fracture occurring in the glass-ceramic although it is possible that yield in the metal component can initiate the fracture. Stresses in glass-ceramic–metal joins are discussed later in this chapter.

3.3.3 Coatings

The use of glass-ceramic coatings on various metals and alloys to provide insulation and/or corrosion resistance is becoming established in a number of different applications. Generally, the glass-ceramic is applied as a powder and bonded directly to the metallic base and is thus a 'special case' of the direct bond discussed above. The metal component is prepared as described above, with care being taken to avoid deformation of the metal parts especially when in thin section. Selection of the metal or alloy to be used depends on the application, and examples of these are referred to below.

There are a number of methods by which the glass powder can be applied to the pretreated metal to provide a powder coating. These are as follows.

1. Spraying – a powder suspension is applied using a compressed gas. This method has been automated in a number of industries, e.g. in the vitreous enamelling of domestic appliances, and automatic processes could equally well be used with glass-ceramics. The process allows thick coatings to be built up, but precise control of thickness can be difficult. This method can be used for coating large areas.
2. Electrostatic deposition – this involves application of a dry powder, to which an electrostatic charge has been applied, onto the metal which is oppositely charged, or earthed, to attract the charged particles. The method is used in similar areas to wet suspension spraying.
3. Screen printing – a powder glass suspension of relatively high viscosity is applied through a wire mesh screen onto the metal. Good definition of areas to be coated can be achieved, leaving unprinted areas such as holes or developing a required pattern. Good control of thickness can be achieved on flat or mono-curved surfaces.
4. Electrophoretic deposition – the application of an electric field attracts charged glass particles in a powder suspension to the substrate. This method permits good control of thickness, permits large areas to be coated and can be used to provide insulating coatings through holes in the metal and to coat irregular shapes.

The powder coatings are allowed to dry, where appropriate, and are then fired, in a controlled atmosphere as necessary (to protect metal parts from undue oxidation), to cause the glass powder particles to coalesce and wet and bond to the metal surface providing the desired continuous insulating layer. Control of the powder particle size distribution (achieved by milling) and of the firing conditions enables coatings to be obtained with very low bubble counts. The firing process, particularly for fairly small samples, lends itself to the use of a continuous-throughput furnace. In some cases the coating is still largely glassy and has to be subjected to a subsequent heat-treatment cycle, essentially as shown in Fig. 3.1, to develop the coating as a glass-ceramic. With certain ranges of compositions, however, the coating will crystallize during the firing cycle, and this is clearly to be preferred from an economic point of view. The development of the crystal phases in the glass-ceramic, which is a process involving diffusion of ions within the glass, can compete and interfere with the diffusion processes involved in the development of the bond and vice versa. As a consequence the crystal phases developing in the glass-ceramic adjacent to the interface may be markedly different from those developing in the coating far from the interface. This is further discussed later in this chapter, and it is obviously important that control of the overall process be achieved and maintained. The advantage of the glass-ceramic coating over a glass coating or vitreous enamel lies particularly in the more refractory nature of the crystalline material. For example, this enables the coating to withstand the firing temperatures used for application of thick-film circuitry which requires firing temperatures in the region of 850–900 °C. Figure 3.14 illustrates glass-ceramic coatings on a Cu–Invar–Cu sandwich, 17%Cr–Fe and titanium onto which thick-film circuitry has subsequently been screen printed and fired at temperatures of 850–900 °C without disturbance of the glass-ceramic layer. In these cases the coatings fulfil the requirements of matching thermal expansion with the particular metallic substrates and are otherwise constituted to provide desired electrical and

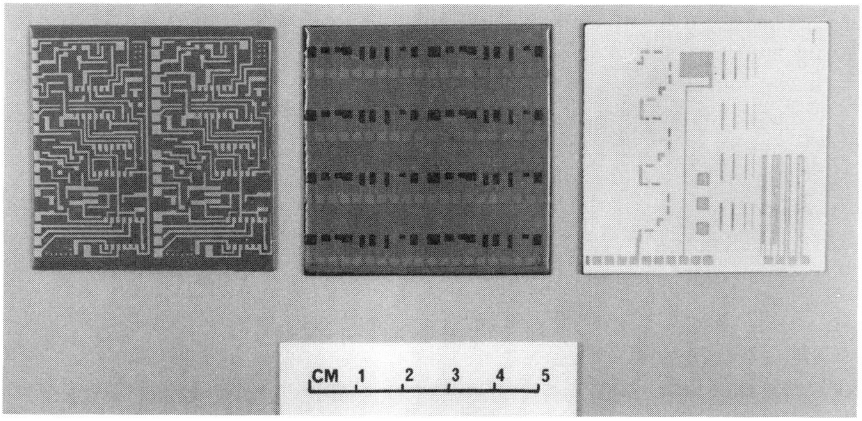

Fig. 3.14 Glass-ceramic coatings on a Cu–Invar–Cu sandwich, 17% Cr–Fe and titanium with screen printed and fired-on conductor and resistor elements.

thermal characteristics [12]. The coatings are usually 50–75 μm thick.

In another example, glass-ceramic coatings have been developed which possess high resistance to corrosion in chemical plant [13]. The glass-ceramic coatings are approximately 1mm thick and are applied onto a refractory glass layer which provides the adhesion to the underlying mild steel. In this application the glass-ceramic coating is deliberately mismatched in expansion to the steel so that compressive stresses are generated in the coating on cooling. This enhances mechanical capability and the ability to resist thermal shock.

3.4 BOND DEVELOPMENT

In order for a satisfactory join to be achieved between the glass-ceramic and metal it is essential that a continuum be formed which successfully bridges the transition between the electronic structure of the metal and the covalent structure of the glass-ceramic. In the various cases discussed previously this linkage is provided via the oxide of the metal. Accordingly it is often advantageous to pre-oxidize the metal parts prior to the joining operation so that oxide is present when the glass comes into intimate contact with the metal. It is important that the correct oxide is developed because, it will be appreciated, not all oxides adhere well to the underlying metal from which they are derived. For example, cupric oxide (black) does not adhere well to copper, whereas cuprous oxide (red) is highly adherent to the metal. In general terms, it is a requisite for the development of good glass-ceramic–metal bonding that a lower-valency oxide of the metal be produced. Thus excessive oxidation of the metal must be avoided during bond development, and hence it is necessary in many instances to carry out the firings in protective atmospheres with low oxygen contents. In brazing operations, where an active metal braze is used in which the titanium content serves to provide the required transition from glass-ceramic to metal, vacuum firing is normally employed. In the other cases an inert atmosphere such as nitrogen or argon provides the protection for the metal component (and graphite sealing jigs if these are employed). In a few cases where a refractory oxidation-resistant metal is used, firings can be carried out in air (an example is the glass-ceramic to 17%Cr–Fe assembly described earlier).

The glass-ceramic, or glass if used in the initial form, must not only react with the oxide formed but must also wet the metal oxide surface in order to permit desired reactions to occur and the required bond to be developed (glass–metal and glass–metal oxide wetting and reactivity behaviour are also discussed in Chapter 2). The capability of a glass, glass-ceramic or metal to wet another surface under the required bonding conditions can be rapidly assessed using the sessile drop technique. In this method, a small piece of the material required to wet, with one angle prepared as a right

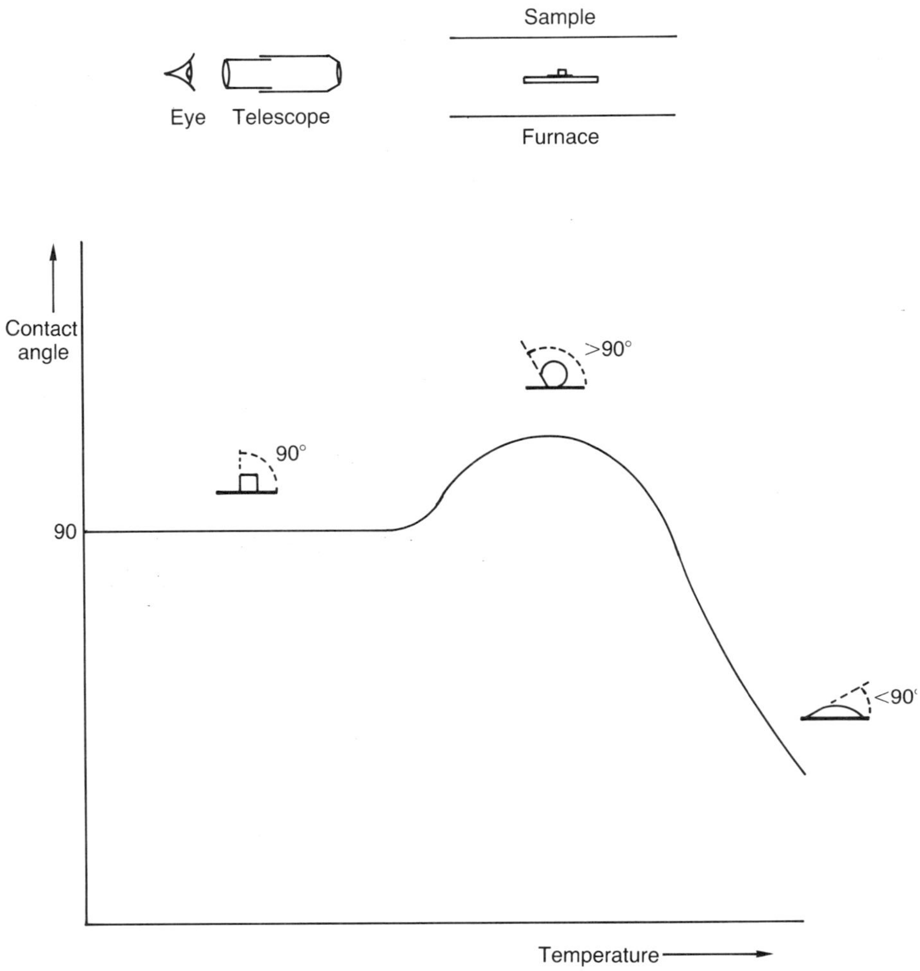

Fig. 3.15 Determination of wetting by the sessile drop method.

angle, is placed on the substrate to be wetted, as illustrated in Fig. 3.15. As the temperature is raised the piece of material rounds to form a 'drop' with its internal angle of contact greater than 90°. As the temperature increases the angle of contact decreases below 90° and the material is then wetting the substrate. A further increase in temperature will decrease the angle of contact further, and the lower the angle the greater is the degree of wetting which is occurring. Of course, it is important that the correct atmospheric conditions for the metal are employed. The information obtained from this experiment can then be used to determine the temperature to be used in bond development for a given combination of glass-ceramic and metal. It may not always be desirable to use temperatures at which contact angles are very low since these may be detrimental to the desired metallurgical characteristics of the metallic element in the system. However, once the conditions are such that the contact angle has decreased below 90° and wetting has commenced, then the components of the system can be brought into intimate contact by the application of force, e.g. by loading on the jigging system or the application of centrifugal forces.

As the bonding reaction proceeds, ions diffuse from the metal into the glass layer. The progress of this diffusion can be followed by examination of samples using scanning electron microscopy (SEM) together with an energy dispersive system (EDS) or investigation by means of electron probe microanalysis (EPMA). This will reveal the crystal structures developing at the interface and the extent to which ions initially derived from the metallic components have diffused into the glass or glass-ceramic layer [14,15].

The bonding which results can be represented by the schematic diagrams shown in Fig. 3.16 [16]. Good bond development is exemplified by the development of covalent bonding via the oxygen ions in the oxide layer and into the glass-ceramic. As noted previously, the presence of a low-valent

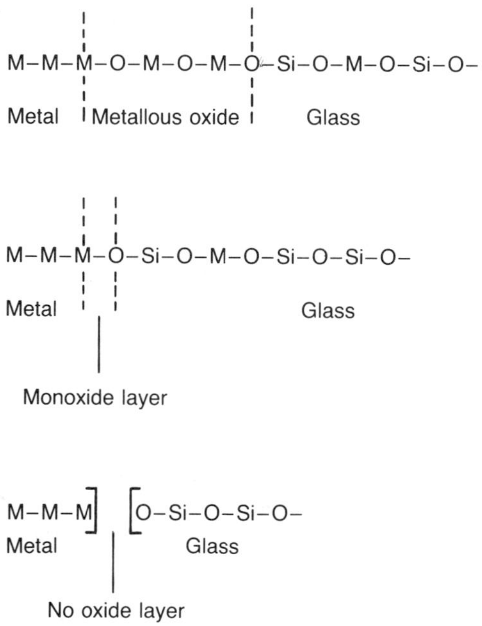

Fig. 3.16 Schematic development of glass–metal boundaries. (Reproduced by permission of the American Ceramic Society.)

oxide (cuprous, ferrous etc.) on the metal is necessary to achieve the desired bond. If insufficient oxide is present, bonding is poor and adhesion is only by means of weak van der Waals forces.

The diffusion of ions in the direct bond situation leads to the occurence of significant changes in composition in the glass adjacent to the interface. The consequence of this is that the crystallization occurring at or near the interface is likely to be very different from that occurring in the bulk glass-ceramic further from the join. Various investigations have been and are being carried out to investigate the interfacial reactions and further to determine the consequences of this variation in crystal content and structure. For example, in joining a glass-ceramic to steel [15] the crystals developing in the bulk of the glass-ceramic are lithium disilicate and quartz. However, owing to the diffusion of Fe ions into the glass-ceramic, a lithium iron silicate crystal phase is developed at the interface.

Although the development of different crystal phases in the glass-ceramic close to the interface is inevitable, and indeed may be desirable to develop the optimum bond, nevertheless problems can result. One problem is that the development of different crystal phases means that the thermal expansion characteristics of the glass-ceramic at the interface are likely to be different from those of the bulk. A requirement for the majority of joints is that the thermal expansion characteristics of the bulk glass-ceramic and the metal are similar. If, however, the thermal expansion characteristics of the interfacial area are very different from those of the bulk material, clearly this may result in the development of high stresses in the interfacial region which may result in failure of the joint [17].

3.5 BOND QUALITY

In general terms bond quality is considered to be satisfactory when the strength of the joint is greater than that of the glass-ceramic component. However, in a number of instances this simplistic view is influenced by seal design and by residual stresses, particularly in the interfacial regions. Consequently, when failure occurs the fracture path is partly in the glass-ceramic and partly in the interfacial region.

Bond quality can be assessed in a number of ways. First, SEM examination of suitable sections can be carried out. However, the information that this imparts, whilst valuable in its own right, depends on other observations carried out first to determine whether a bond has high strength or not. Consequently, SEM investigation is particularly useful as a means of determining why a bond is good or poor but not as a way of making the initial decision.

Secondly, the tensile strength of the joint can be determined. In the case of joined bulk components this usually requires the preparation of special test pieces such as that illustrated in Fig. 3.8. The strength of the joint can then be determined by tensile pull, taking care to ensure proper alignment during the test. In the case of coatings on metal, circular pads are screened onto the metal and fired on. Tensile pull metal pieces are then bonded to each side of the glass-ceramic–metal composite using a suitable epoxy resin, after which the tensile strength of the bond can be determined [18]. Alternatively, the glass-ceramic–metal composite can be subjected to shear stressing until failure occurs. The preferred test specimen is rectangular in section in order to ensure uniform loading across the bond line. Bending strength measurements in three- or four-point loading can also be used and give comparable results to those of tensile tests [19].

Thirdly, the stress in the joint can be determined. Because the glass-ceramics employed are generally opaque in the visible, it is not possible to use the photoelastic techniques which are commonly used to examine glass–metal joints. It may be possible to carry out similar measurements in the infrared where many glass-ceramics have at least some transparency. However, development of infrared photoelastic techniques is at an early stage and, whilst these may offer some possibility for the future, they are not currently available.

Another technique relying on transmission is based on X-ray diffraction (XRD). Crystal lattices under stress distort and the degree of this distortion can be determined by XRD techniques. However, this method is only likely to be useful in the case of intermediate bonds where the crystal morphology of the bulk glass-ceramic is already established so that changes due to stress close to the interface can be determined. In the case of direct bonding, the crystal structure close to the interface is determined by the extent of the reaction during bonding procedures and does not represent a change due to stress. Of course, the interlayer is likely to be stressed, but unless the precise details of the crystal structure at the interface can be determined in the unstressed condition structural distortion and hence stress levels cannot readily be established.

A possibility for direct determination of stress involves the use of a bending beam technique. This method has been used successfully with different thicknesses of glass-ceramic layer to establish the influence of compositional change due to reaction on stress at the interface [17] and can easily be extended to enable stress in a variety of glass-ceramic–metal combinations to be established.

However, the residual stress in a seal assembly is influenced not only by differences in thermal expansion characteristics and other parameters such as elastic moduli but also by the design of the seal. Under these circumstances the direct measurement of stress by the bending beam method may not be totally applicable. Consequently other means of arriving at a value of the stress in the joint must be adopted. One technique is to employ finite-element analysis methods [20]. Such methods have been shown to give an accurate picture of stresses in joints where these could also be measured by photoelastic techniques, and this has given confidence in the use of finite-element techniques for stress evaluation.

3.6 CONCLUSIONS

The preparation of a number of glass-ceramic–metal joints using intermediate bonding media or directly joining the glass-ceramic to the metal has been described. The method selected depends on the design of the seal and, to some extent, on process and service conditions. Methods of establishing seal quality and measuring the stress in the joint have been described. Clearly, this is a technology which is continually developing, with the preparation of new combinations and the development of an understanding of interfacial reactions and the influence of stress on bond quality. The discussion in the preceding sections serves to provide the reader with general concepts and to give some indication of the direction in which further information might be obtained.

ACKNOWLEDGEMENTS

The author thanks GEC ALSTHOM Ltd for permission to publish this paper. He would also like to acknowledge the contributions of his colleagues at the GEC ALSTHOM Engineering Research Centre, in particular A.K. Datta, C.A. Elyard and T. Hales.

REFERENCES

1. de Réaumur, M. (1739) Art de faire une nouvelle espèce de porcelaine, *Mem. Acad. Sci.*, 370–88.
2. Stookey, S.D. (1956) Corning Glass Works, BP 752, 243, 11.7.56.
3. McMillan, P.W. (1979) *Glass-ceramics*, 2nd edn, Academic Press, London.
4. Strnad, Z. (1986) *Glass-ceramic Materials*, *Glass Science and Technology 8*, Elsevier, Amsterdam.
5. Partridge, G. (1982) Nucleation and crystallization phenomena in low expansion $Li_2O-Al_2O_3-SiO_2$ glass-ceramics, *Glass Technol.*, **23** (3), 133–8.
6. Rabinovich, E.M. (1982) Cordierite glass-ceramics produced by sintering, in *Nucleations and Crystallization in Glasses. Advances in Ceramics*, Vol. 4 (eds J.H. Simmons, D.R. Uhlman and G.H. Beal), American Ceramic Society, Columbus, OH.
7. Clark, T.J. and Reed, J.S. (1986) Kinetic processes involved in the sintering and crystallization of glass powders, *J. Am. Ceram. Soc.*, **69** (11), 837–46.
8. Mizuhura, H. and Huebel, E. (1986) Joining ceramic to metal with ductile active filler metal, *Weld. J.*, **65** (10), 43–51.
9. Partridge, G. and Phillips, S.V. (1978) New developments in glasses and ceramics for the electrical industry, *GEC J. Sci. Technol.*, **45** (1), 11–20.
10. McMillan, P.W., Hodgson, B.P. and Partridge, G. (1966) Sealing glass-ceramics to metals, Part 1, Selection of materials and direct sealing methods, *Glass Technol.*, **7** (4), 121–7.

11. Partridge, G. (1985) The use of fibre viscometry to study nucleation and crystallization phenomena in glass-ceramic compositions, *Glass Technol.*, **26** (3), 129–33.
12. Partridge, G., Elyard, C.A. and Budd, M.I. (1989) Glass-ceramics in substrate applications, in *Glasses and Glass-ceramics* (ed. M.H. Lewis), pp. 226–71, Chapman and Hall, London.
13. Partridge, G. (1981) Improved enamels for chemical plant, *GEC J. Sci. Technol.*, **47** (2), 87–94.
14. Partridge, G. and Elyard, C.A. (1984) Glass-ceramic bonds to metals, *Proc. Br. Ceram. Soc.*, **34**, 219–29.
15. Partridge, G., Elyard, G.A., Logan, E.A. and Holland, D. (1989) Glass-ceramic bonds to metals, in *Joining Ceramics, Glass and Metal* (ed. W. Kraft), pp. 163–70, Oberusel DGM.
16. Pask, J. (1971) Chemical reactions and adherence at glass–metal interfaces, *Pro. Porcelain Enamel Inst. Tech. Forum*, **33**, 1–16.
17. King, S.C. and Loehman, R.E. (1987) Thermal expansion mismatch produced by interfacial reactions in glass-ceramic to metal seals, *Adv. Ceram. Mater.*, **2** (1), 69–73.
18. Sturgeon, A.J. *Glass-ceramic Coatings on Metals*, Ph.D. Thesis, University of Warwick.
19. Suganuma, K., Okamoto, T., Koizumi M., and Shimada, M. (1989) Relationship between the strength of ceramic/metal joints in tensile and three-point bending, *J. Am. Ceram. Soc.*, **69** (10), C235–6.
20. Varshneya, A.K. and Petti, R.J. (1978) Finite element analysis of stresses in glass-to-metal foil seals, *J. Am. Ceram. Soc.*, **61** (11–12), 498–503.

4

Metallizing ceramic surfaces

A.W. Hey

4.1 INTRODUCTION

It is difficult to form joints by brazing ceramics, particularly oxide ceramics, because few commercially available brazes can wet their surfaces. This problem is usually overcome by treating the ceramic surfaces to give them metallic wettable characteristics. For example, metal coatings can be applied by vapour deposition or sputter ion plating. However, the most widely used metallization process for oxide ceramics is that in which mixtures of a glassy phase and a refractory metal are applied as paint by brushing or screen printing, as exemplified by the 'moly-manganese' process. The prime purpose of the glassy phase is to bond the refractory metal to the ceramic oxide, while that of the refractory metal is to render the surface amenable to electroplating and ultimately brazing or soldering to effect a joint.

Metallization of oxide ceramics followed by plating and brazing has been established for several decades. Therefore the problems involved in its application are those of practice rather than principle, and in this chapter we address practical aspects of metallizing using refractory metals such as tungsten and molybdenum. When received the ceramic will have undergone, in many instances, a very costly fabrication and machining process (Fig. 4.1) and its value will be much higher than that of the starting material. Consequently it is essential to maintain a strict control on quality during the metallizing process as any defect caused by poor conditions or procedures may mean that the part will have to be scrapped.

In the following sections we describe some of the pitfalls commonly encountered and indicate some of the ways in which quality can be maintained in a reproducible and reliable product. Successful joining of metallized ceramics depends on optimization of each of the three or four stages of the process relating to the selection and preparation of the ceramic, primary metallization with glass–refractory metal mixes, secondary metallization, when required, by plating, and brazing.

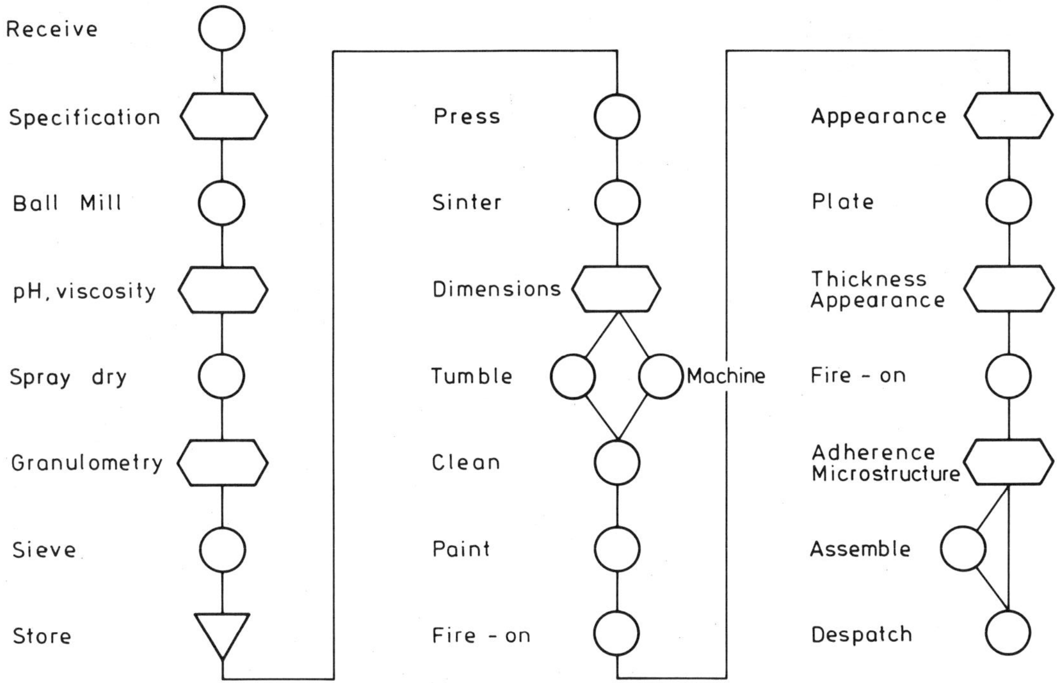

Fig. 4.1 Generalized process flowchart for ceramic fabrication and metallizing.

4.2 THE CERAMIC

4.2.1 Chemistry

The ceramics to be metallized have progressed from impure debased oxide ceramics with high silica contents to pure materials such as beryllia and alumina, which now account for the majority of metallized components in electronic devices. To metallize these oxides it is necessary to use strictly controlled conditions, which do not always lend themselves to scaled-up industrial production unless debasing additions are made to the ceramics. In selecting the ceramic and the debasing addition, care must be exercised to take advantage of the oxide attributes such as a low sintering temperature and to maintain density and grain size without affecting the metallizing character of the product. In some rare instances it is possible to enhance both metallization characteristics and ceramic properties. The application of the metal–ceramic assembly will dictate the choice of the primary ceramic, but the choice of debasing materials can be made from those which produce a liquid phase (CaO, SiO_2), those which have an effect on sintering rates or grain growth (MgO) and those which react at the metal–ceramic interface (TiO_2).

A detailed discussion of our understanding of these effects is given by Reed *et al.* [1], and many proprietary formulations now exist which embody the general principles enunciated by these workers.

4.2.2 Surface finish

The finishes normally demanded for metallized components can be met by 'as-fired' and machined surfaces. Some of the defects of 'as-fired' ceramics are shown in Fig. 4.2. Machined, lapped or ground surfaces may also have some disadvantages: they are more costly to produce; the surface chemistry is different from that of the fired product and this could degrade the metallizability; the finish can be rough because of poor machining practice or surface pull-outs; the surface is contaminated by the machining tools and must be cleaned, often at considerable cost.

Almost all fired ceramics are pressed from powder and can have faults caused by worn or poorly set up tooling. It is essential to remove any flashing or protrusions by tumbling before attempting to metallize. Even slight flashing can cause chaos in automatic handling machines, leaving areas which are not metallized, cutting through screens, misaligning jigs and causing gaps in assemblies which cannot be filled with braze.

4.2.3 Surface cleanliness

If it is necessary to machine surfaces it is vital that *all* surfaces are subsequently cleaned to the highest standard to remove grease from

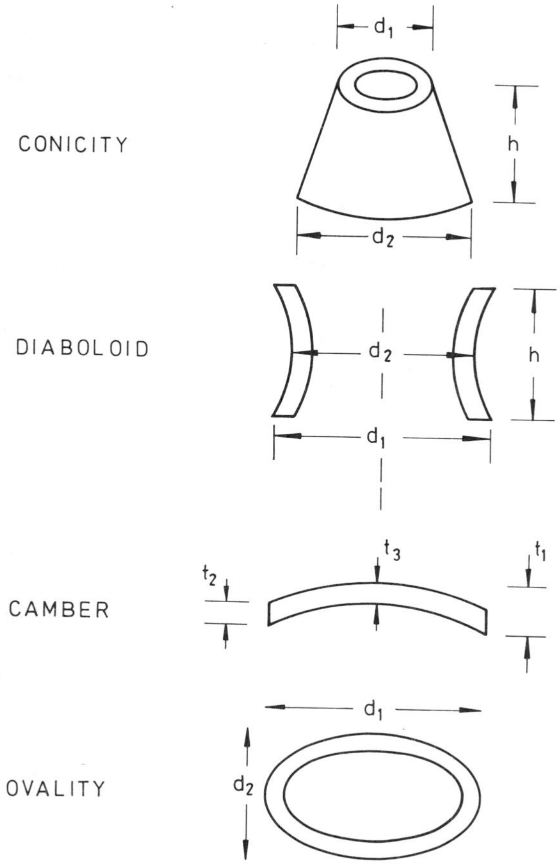

Fig. 4.2 Some critical tolerances for as-fired ceramic.

handling or from cutting fluids, abrasion debris, metallic contamination from grinding wheels, and water.

Stringent processing demands the use of hot and cold water rinses with surfactants, sometimes acid or alkali baths to remove surface contamination, ultrasonic baths to eliminate particulate matter from pores, and a final rinse in a solvent followed by hot air drying. No surfaces should be in contact at the final drying stage. Once all the 'dirt' has been removed, a frequent final treatment is to 'anneal' the surface by heating in air at a temperature of 800–1000 °C.

4.3 THE PRIMARY METALLIZING LAYER

The primary metallizing layer is the mixture of refractory metal and glass which is coated onto the surface of the ceramic. The most commonly used material is a 'paint' containing molybdenum and manganese. An alternative choice of refractory metal is tungsten, which is used when gold is to be plated onto the surface without an intermediate layer of nickel.

4.3.1 The metallizing paint

The paint is an intimate blend of a refractory metal (tungsten or molybdenum), manganese oxide when the metal is molybdenum, a glass frit, a carrying vehicle and a solvent. Typical formulations are given in references 1 and 2. Commercially manufactured paints are available in the United States, but most UK companies with metallizing plant prepare their own paints based on in-house recipes. In achieving the best possible metallizing many parameters need to be considered, of which the most important are the application method and the process parameters, particularly the temperature, time, atmospheric conditions and the thickness of the paint.

There are several ways of applying the paint to the ceramic ranging from simple manual dipping to fully automated machines. Among the more popular methods are spraying, screen printing, syringe or nozzle brushing and transfer using a wheel or tape. The application method chosen is largely determined by the shape of the article, the complexity of the metallizing pattern and the quantity or quality desired. No matter which method is chosen it is essential to ensure that the paint layer is of the correct thickness, usually at least five times the molybdenum or tungsten particle size which is commonly about 5 μm. The thickness achieved in practice depends on both the paint and the method of application. One of the most important paint characteristics is its viscosity. A paint with no thixotropy will flow after application and definition or precision will be lost. Additionally, the coating may be too thin after the firing process. A paint which requires a high shear stress to make it flow may damage screens or make nozzle application impossible. Ideally, a paint with a low yield point which sets immediately after the shear stress is relaxed will avoid such defects as screen impressions, uneven thickness or bubbles [3]. Paint thicknesses of less than 5 μm can result in braze penetration during subsequent assembly operations, while thicknesses greater than 5 μm may cause cracks or blisters during drying which may result in leaks or peeling of the metallizing. Defects which can be caused by an incorrect paint thickness are shown in Fig. 4.3.

4.3.2 Firing process variables

During the firing cycle the metallizing paint undergoes many changes. At an early stage there is decomposition and elimination of organic materials in the paint. Later, and even more importantly, there is sintering of the

Method of ink application	Form of dried/fired metallising profile	Thickness (μm)	Quality
Spraying	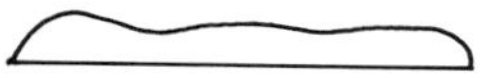	0–50	– Uneven thickness – No precision
Screen Printing		10–40	– Mesh lines – Very precise
Nozzle		5–40	– Hump back, i.e. variable thickness – Can be very precise – Influenced by precision of ceramic
Brush		0–50	– Brush marks – Uneven thickness – Limited precision
Wheel		5–30	– Variable width and thickness as given by ceramic quality – Precise – Flat profile
Tape		5–20	– Dependant on tape quality

Fig. 4.3 Form (profile) and quality aspects of metallizing in relation to the method of application.

refractory metal particles and diffusion of the glassy phase from the paint onto and into the ceramic or vice versa. Figure 4.4 shows some of the effects of temperature variation on the structure of the metallized layer.

The bonding between the ceramic and the metallizing layer is produced by diffusion of a glassy phase from the paint, in the case of high-purity ceramics, or by diffusion of glassy binder phase from the ceramic into glass-free paint on the surface of a debased ceramic. In ideal conditions, glass covers the surface of the ceramic or fills the pores of the refractory metal or both [4]. Because the diffusion process is both temperature and time dependent, the integrity of the structure can be degraded (Fig. 4.4) by the following mechanisms:

1. glass diffusion from the paint into the ceramic which denudes the metallizing layer and allows sintering of the metal particles to produce a fragile crust in the structure;

Metallizing conditions	Resulting microstructure	Effects
Temperature too low Short time Excessive humidity or Dry atmosphere		Metallizing friable No bonding Difficult to plate Metallizing structure inconsistent
Optimal	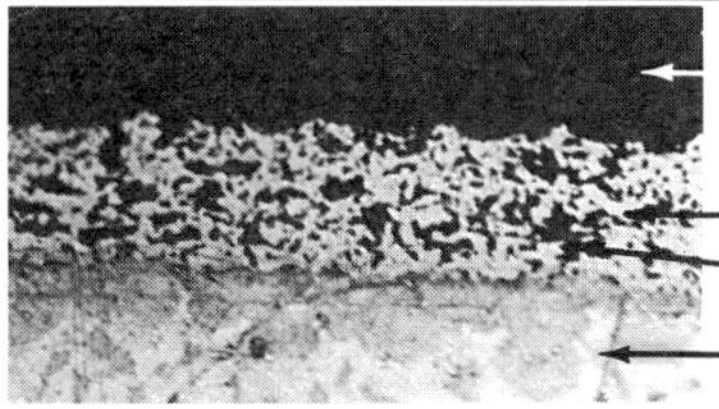	Excellent adherence Bright molybdenum Easy plating Homogenous/coherent structure
Temperature too high Long time Excessive humidity or Dry atmosphere	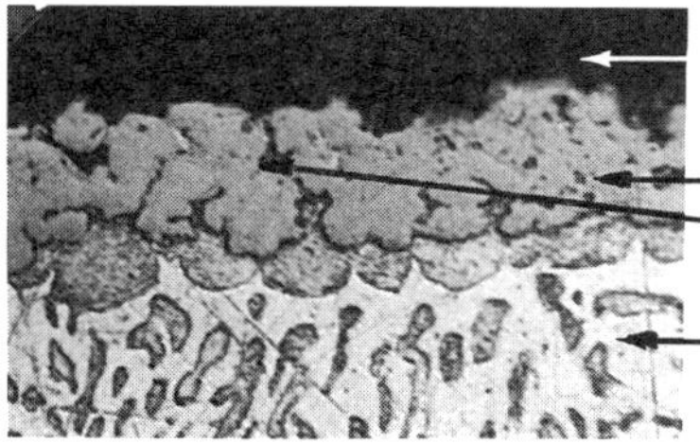	Poor adherence Large Mo grains No glassy phase or an excess of glass on surface Braze penetration Leaks Plating difficulties

Fig. 4.4 Influence of various processing conditions on metallizing.

2. glass diffusion from the ceramic binder phase into the metallization layer to such an extent that the surface is flooded, subsequent plating is no longer possible and brazing cannot be achieved.

Experimental work is essential to define the critical values of time and temperature required to maintain reproducible structures.

The metallizing paint must be heated in a reducing atmosphere if oxidation of the refractory metal is to be avoided. Practical and safe ways of achieving this are available using atmospheres such as pure hydrogen, forming gas ($4N_2 : 1H_2$) or cracked ammonia ($3H_2 : 1N_2$). In general the gas or gas mixture is passed continuously through the furnace and ignited at the exit point.

To ensure that the glass frit wets the surface of the refractory metal well it is almost always necessary to introduce water vapour into the reducing gas [4,5]. How much water vapour is required depends on the local conditions and must be established by experiment. In some practices the water content is controlled by bubbling the incoming gases through water at different temperatures. This can lead to an excessive amount of water in the furnace as there are other reactions occurring which produce water in addition to that added deliberately. An alternative method is to monitor the water content of the outgoing gases and then feed back the information to a system which will bypass the humidifier if excess water is present. Different furnaces give different results, and so once again the preferred moisture level must be determined by experiment.

4.4 THE SECONDARY METALLIZING LAYER

The most commonly used solders and brazes will not wet the surface of the primary metallizing layers. To achieve wetting, a secondary metallizing layer of a metal such as nickel, copper, gold, tin or lead is deposited by plating (electrolytic or electroless). The plating operation contributes as much to the success or failure of the metallizing structure as the primary layer. The process must be controlled throughout and Fig. 4.5 shows a process flow chart for a typical nickel plating line which is adaptable to rack or barrel methods. Flow charts for the other metals commonly used as secondary layers would be similar, but an additional series of baths is necessary for tin or lead plating, as indicated in the figure.

The roles of the plated layer are to aid wetting of the refractory metal by the braze and to resist braze penetration into the primary layer. One key parameter which affects both these roles is the thickness of the plated layer. If it is too thin the braze may dissolve it, but if it is too thick blisters can form during the braze process which lead to very poor adherence and leaks. In general a plating thickness of 2–4 μm is sufficient to avoid such problems.

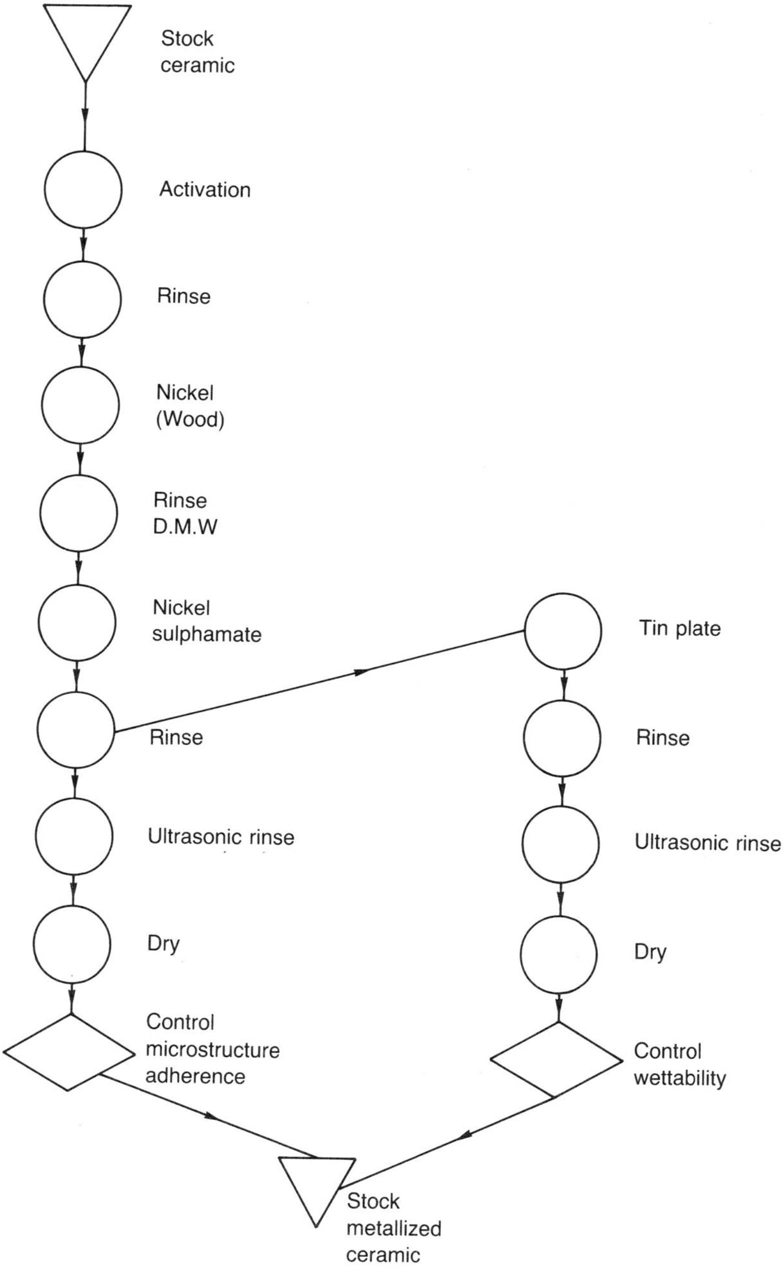

Fig. 4.5 Process flowchart for nickel electroplating.

The barrel method is generally used for plating areas which are very small or are not easily accessible. Contact between the area being plated and the barrelling media (pins, balls etc.) is intermittent, and the plating can be thin and variable. One technique used to increase the thickness and improve the quality of the plating is first to plate 1–2 μm by barrelling the desired metal, then to diffuse this layer into the refractory metal by heating in a reducing atmosphere (in the case of nickel or gold at 800 °C) and finally to replate to give a total thickness of 2–4 μm. The diffusion process can be performed in continuous or batch brazing furnaces. As well as providing a surface which plates more readily, the process allows an opportunity for examination of the plated layer for defects such as blistering.

Polycrystalline ceramic surfaces are almost always porous, and this is particularly true of those which have been machined. Plating salts remaining in pores must be removed by a careful washing operation or they will discolour the ceramic after diffusion or brazing and produce a conducting layer which can degrade the performance of the plated product.

Electroplating baths can be made up according to well-tested recipes or purchased from suppliers who are electroplating specialists. Plating equipment can be very sophisticated, but any equipment can give disastrous results unless quality control is strict, the correct procedures are followed and no short cuts are taken. A mistake or error in the plating operation is the commonest cause of the rejection of metallized components.

Electroless plating is an alternative procedure for some components. It relies on chemical processes to coat the surface of the primary metallizing layer. The simplest process is the deposition of copper onto iron from a copper sulphate solution. Other solutions produce nickel phosphide (Ni_3P) or nickel boride (Ni_2B) coatings. Although there are some advantages such as a low capital outlay, it can be difficult to start the reaction, the solutions can be unstable and plating can occur where it is not required. The coating thickness is difficult to control, and the process requires more analytical control than electroplating does.

4.5 THE BRAZING STAGE

Brazing techniques and their associated problems are discussed elsewhere in this book, but to complete this overview of metallizing practice a brief description of brazing faults and possible remedies is required.

Many structural or chemical faults in the metallizing layers are immediately evident during the brazing operation. Thus glass-rich primary layer surfaces do not plate and brazes cannot wet the surface, so that there is poor adherence. Braze penetration under the primary layer can occur, causing blisters, bubbles and poor alignment. Similarly, thin secondary layers dissolve in the braze which can then penetrate poor primary metallizing layers or de-wet to produce holes which weaken the structure

Fig. 4.6 Poorly brazed structures.

or act as leak paths through hermetic or vacuum seals. Figure 4.6 shows some defects at a low magnification. Examination at higher magnification can give further information about the causes, and thereafter the process can be modified.

4.6 THERMAL CONTRACTION FAULTS

Various faults relating to the ceramic, the primary layer, the secondary layer or the braze have been described in previous sections. There is one further major fault which can occur after brazing – cracking in the ceramic as a result of thermal mismatch between a metal part brazed to a ceramic. This mismatch can cause major problems and some typical faults are

Fig. 4.7 Thermal mismatch cracks in ceramic–metal assemblies.

illustrated in Fig. 4.7. Some of the failures can be attributed to poor design, some to inherently weak ceramics and others to incompatibility of the ceramic and the metal. It is not our purpose in this chapter to deal with design but rather to indicate how to avoid situations which may contribute to failure.

Fig. 4.8 Brazed assembly showing the influence of rounded (tumbled) ceramic.

In an earlier section it was pointed out that tumbling ceramics removes the flashes produced during the ceramic powder pressing operation. Whilst flash removal is essential to avoid some of the pitfalls when applying paint, the rounding of sharp angles has other benefits, especially in the brazed component (Fig. 4.8). Inherent stresses along the surface caused by mismatched contraction on cooling can be accommodated better by a rounded surface and it is generally accepted practice to chamfer corners to reduce the risk of cracking. If design restrictions do not allow this, low thermal expansion metals which more closely match the coefficient of the ceramics must be used. Large masses of high thermal expansion metals such as copper should be avoided in favour of thin metal components of low thermal expansion nickel or Fe–Ni alloys.

4.7 TESTING METALLIZED COMPONENTS

The testing of metallized components is probably the most contentious subject in this chapter. Strength testing is dealt with in Chapter 8, but it is worth briefly referring to some of the more basic methods used to detect faults.

4.7.1 Microstructure

Metallographic cross-sections (Figs 4.6–4.8) of metallized/plated components can give valuable insight into controllable factors such as the thickness of the refractory (primary) metal layer, the secondary metal (plated) layer, uniformity, braze penetration, de-wetting, and crack formation. Although such an examination is subjective to some extent, it can complete the critical appraisal of the quality of the metallizing structure when the results of other tests are known, and in many instances is the only way to determine what has failed in the metallizing process. Microstructural examination should be considered as a necessary routine procedure and not one which is used only for troubleshooting. A data bank of microstructure covering a time span can indicate minor changes in process or material conditions which can be easily missed in infrequent random checks.

4.7.2 Adherence

No single adherence test is sufficient to judge the quality of a ceramic–metal seal. For example, it is possible to have a tensile strength well within specified limits while the same part (e.g. a lead frame on a ceramic package) will fail readily when subjected to a peel test. Both tests are essential to assess the quality of the metallizing and together with the microstructure they can adequately define the overall quality of the process.

The manner in which the ceramic–metal interface ruptures can give vital clues about which stage of the process may be defective. Fracture within the ceramic and strengths within the specification limits are ideal in the sense that the ceramic is weaker than the seal. However, interfacial failure can give clues to the quality of the metallizing and plating techniques. Thus it is possible for a metallizing layer to be stripped away completely from the ceramic face, and for the secondary layer to pull away from the primary layer.

The two main test methods available for assessing tensile strength are shown in Fig. 4.9. Both aim at reducing effects such as misalignment which

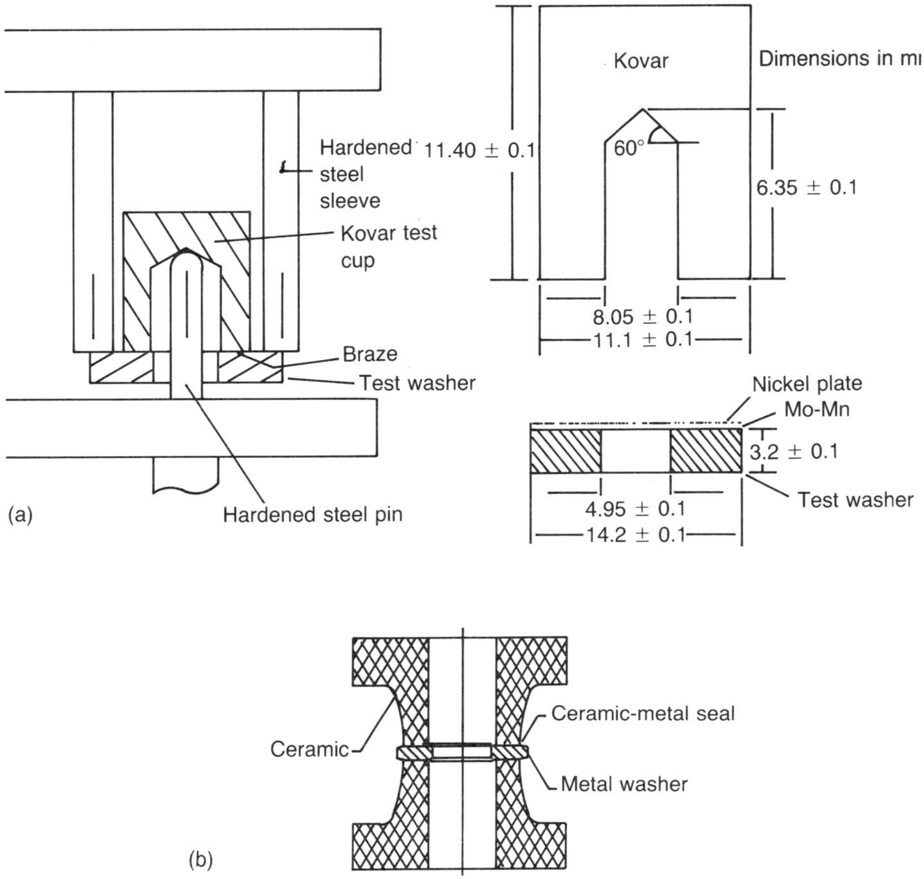

Fig. 4.9 Tensile test assemblies: (a) Kovar cup and ceramic washer test; (b) ASTM test.

can influence the measured strength of the seal. There are advantages and disadvantages with any method, but if the test results are reliable from one test batch to another the technique can be regarded as adequate. Criteria such as reproducibility, reliability and fitness for use are more important than absolute value to a quality manager faced with decisions on process control.

4.7.3 Visual examination

During each stage of manufacture, defects can occur which in themselves do not affect the properties of the metallizing. However, they can be the cause of high rejection rates. Visual examination is essential many times

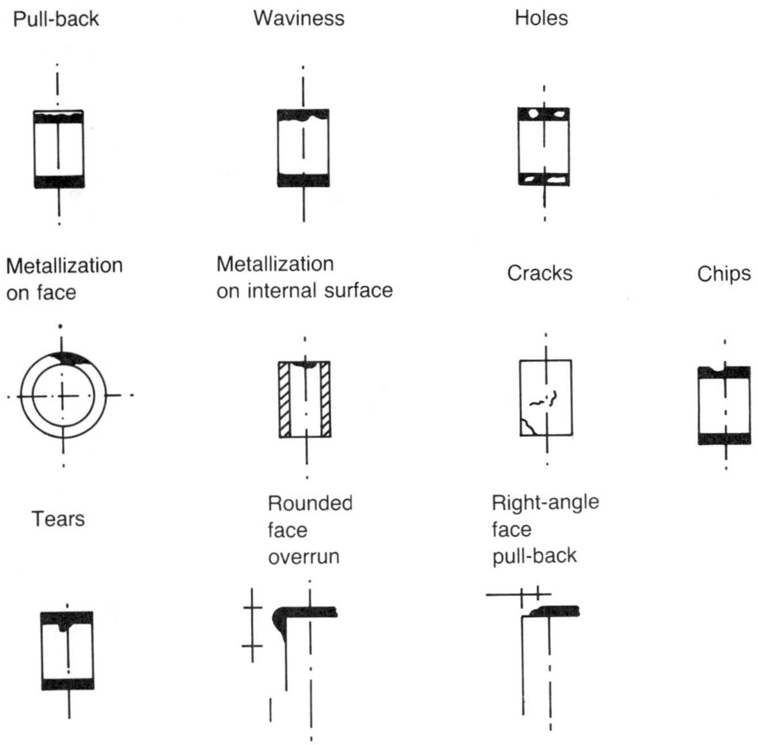

Fig. 4.10 Faults contributing to the poor visual quality of metallized ceramics.

during the process because it affords a quick guide to the behaviour of the process. Fig. 4.10 illustrates some of the defects which cannot be accepted on many components. Those cited are not exhaustive and are process dependent. Others, caused by bad operator habits, include fingerprints, paint splashes, tweezer marks and dust or hair, but there are many others.

The refractory metallizing process is often referred to as a black art, and a vigilant eye, good housekeeping and well-tested procedures are vital to complement the physics, chemistry and metallurgy which form an integral part of the equation of successful metallizing.

4.8 CONCLUSIONS

The metallization of ceramics by the refractory metal process is long established. It is a complex mixture of many sciences which combine to give the desired result: the technology is full of pitfalls, but these can be avoided. The manufacture and properties of the ceramic, and the composition of the metallizing paint, its application, subsequent firing and further treatment, all require strict adherence to procedural discipline. Minor variations from the established process conditions and procedures can have disastrous effects on the quality of the product. The combination of an empirical approach to problem solving together with an understanding of the scientific principles involved at each process stage leads to a viable manufacturing route. Coupling this to statistical process control and other technical back-up facilities reduces the risk of errors being made and ensures product reliability.

REFERENCES

1. Reed, L., Wade, W., Vogel, S., McRae, R. and Barnes, C. (1966) *AD636950, Report No 12, TR66–1, R/E 66–115*, Metallurgical Research and Development for Ceramic Electron Devices (Elmac), Division of Varian.
2. Kohl, W.H. (1967) *Handbook of Materials and Techniques for Vacuum Devices*, p. 441, Reinhold, New York.
3. Holmes, P.J. and Loadsby, R.G. (1976) *Handbook of Thick Film Technology*, Electrochemical Publications, Ayr, Scotland.
4. Twentyman, M.E. (1975) *J. Mater. Sci.*, **10**, 765, 779, 791.
5. Helgessen, C.I. (1968) *Ceramic to Metal Seals of Heavy Metals to Beryllia Ceramics*, pp. 28, 109, Boston Technical Publishers, Cambridge, MA.

+C
Ru + SiC → RuC & RuSi

Ru + C → RuC

V + SiC → VSi + C + VC

V + C → VC

Pt + SiC → PtC + PtS

Pt +

5

Active metal brazing

M.G. Nicholas

5.1 INTRODUCTION

Ceramic workpieces can be directly brazed to other ceramics and to metals without prior surface coating. However, the range of brazing metals and alloys that can be used and the joining conditions employed are far more restricted than for metal–metal joining. A few carbide ceramics can be brazed using metals and alloys employed originally for metal joining – thus silver alloys can be used to braze WC–Co cutting tool tips to steel shanks [1] – but the joining of most ceramics requires the use of special active metal brazes.

Many technically important ceramics, including alumina, zirconia, beryllia, silicon carbide, titanium carbide, silicon nitride, boron nitride and sialons, are not wetted by silver, copper or gold – the basic constituents of most brazing alloys. That is, the molten metals will not spread over the ceramic surfaces or flow into capillary gaps. The conventional measure of wettability is the contact angle, as illustrated in Chapter 2, Fig. 2.5, for sessile drops resting on horizontal substrates, which can be related to the surface and interfacial energies, or tensions, of the system following the argument of Young [2]. Mathematically, this can be expressed by the equation

$$\gamma_{SV} = \gamma_{SL} + \gamma_{LV} \cos\theta \qquad (5.1)$$

where γ is the energy, or its numerical equivalent tension where the units are $J\,m^{-2}$ and $N\,m^{-1}$ respectively, and the subscripts identify the solid–vapour and liquid–vapour surfaces and the solid–liquid interface. In the case of a liquid metal drop in contact with a solid ceramic, equation (5.1) can also be written as

$$\gamma_c = \gamma_m \cos\theta + \gamma_{cm} \qquad (5.2)$$

If the interfacial energy γ_{cm} is larger than the ceramic surface energy γ_c, the liquid will not spread freely or flow into capillary gaps, $\cos\theta$ will be negative and the contact angle will be greater than 90°.

The reason for this general lack of wettability is that the ionic or covalent structures of ceramics lack the delocalized bonding electrons present in metals [3] and hence ceramic–metal interfaces are major electronic discontinuities. Therefore, since non-wetting is associated with the basic chemistry of ceramic–metal interfaces, it can be argued that wetting might be promoted if the braze reacted to change the surface chemistry of the ceramic. However, the more chemically stable ceramics are generally the least wetted [4], and hence the interfacial reactions will usually be detrimental. Fortunately there are exceptions to this, as to all other generalizations, and these exceptions are exploited in the active brazing process.

5.2 BRAZES

5.2.1 Range and availability

The patent and technical literature advocates the virtues of a very wide range of formulations as active metal brazes. At first sight it seems that almost every conceivable combination of materials has been suggested by somebody. Even a cursory glance reveals that chromium, hafnium, lanthanum, tantalum, thorium, titanium, tungsten, vanadium and zirconium have been advanced as active metal additions to brazes based on or containing large proportions of beryllium, cobalt, chromium, copper, gold, iron, lead, manganese, molybdenum, nickel, niobium, platinum, silver, tellurium or tin. Most suggestions, however, relate to modifications of conventional braze alloys based on copper, gold, nickel or silver by the addition of titanium or zirconium.

As is usually the case with patent literature, little general guidance is given about the behaviour or characteristics of the materials cited. However, technical literature can be useful to those seeking guidance, and it is noteworthy that these descriptions of active metal braze compositions are much more restricted in range than those in patents. The material which has been most thoroughly described is the Ag–28Cu eutectic to which up to 10% of titanium has been added (unless stated otherwise, alloy compositions are given in weight per cent). Recently considerable attention has been paid to the Ag–27.5Cu–2Ti and Ag–24Cu–15In–1.5Ti alloys. Other frequently mentioned alloys include titanium-based materials, in particular the Ti–28Ni eutectic and – by workers from the Oak Ridge National Laboratory in the United states – the Ti–49Cu–2Be, Ti–28Ag–4Be and Ti–49Zr–4Be alloys [5]. Recently, these workers have reported experience with the Ag–24Cu–15In–1.5Ti and Ag–27.5Cu–2Ti alloys and a range of experimental alloys such as Ag–26Cu–4Sn–4Ti [6].

Other workers have also reported that tin and indium are interchangeable, describing work with Cu–11In–21Ti and Cu–15Sn–15Ti alloys and also with Cu–27Au–13Ti and Cu–12.5Ag–25Ti alloys [7]. Finally, the use of pure aluminium as a braze for SiC and some other engineering ceramics has been described by Japanese workers [8,9], and its use with Si_3N_4 is described in Chapter 10.

Although the range of materials suggested by laboratory studies is large, the number of braze alloys that are commercially available is far more restricted. This reflects the small size of the market overall and the unattractiveness of holding stocks of many products that have similar performance. The longest established active metal braze is the Ag–28Cu eutectic to which 5%–10% of titanium has been added mechanically to produce sandwich sheets or cored wires. While the eutectic temperature for the Ag–Cu system is 780 °C, the use temperatures of these mechanical alloys are usually significantly higher because of the need to dissolve the titanium and the non-eutectic character of the ternary alloy. The Ag–27.5Cu–2Ti alloy has become available in recent years and is finding many applications. This alloy melts at 780–795 °C and can be used at 820–860 °C, while the Ag–24Cu–15In–1.5Ti alloy now becoming commercially available melts at 605–755 °C and can be used at 750–800 °C. This quaternary material is described as a ductile active metal braze, permitting preforms to be stamped from sheet, and this is also characteristic of the Cu–Ti alloys now being produced experimentally using rapid solidification techniques. The Ti–28Ni eutectic braze alloy, which melts at 955 °C, is not normally available as a commercial material but has been used by a number of workers as, for example, nickel-plated titanium wire preforms or rapidly solidified sheet.

Perhaps the most notable point about these braze compositions is the dominant role of titanium as the active agent. In fact there are grounds for arguing that titanium is an advantageous additive because of the nature of its reactions with ceramics.

5.2.2 The role of titanium in brazing oxides

Titanium can dissolve 29 at.% of oxygen at 1125 °C and forms a family of oxides ranging from TiO to TiO_2 (Fig. 5.1) [10]. Thus in order to understand the role of titanium in the active metal brazing of oxides it is important to know which oxide will be formed as a product by reactions such as

$$3[Ti] + Al_2O_3 \rightleftarrows 2[Al] + 3TiO \qquad (5.3)$$

in which square brackets identify solutes. If the product is assumed to be stoichiometric, the energy change ΔG_R associated with the reaction, can

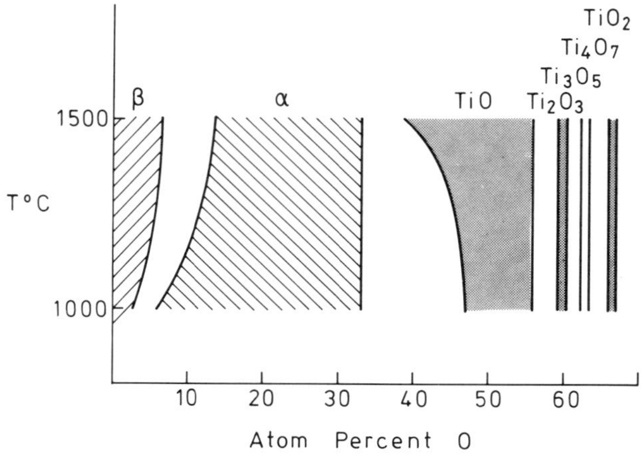

Fig. 5.1 Ti–O phase diagram [10].

be related to the free energies of formation of the oxides and the metal activities by the equations

$$\Delta G_R = 3\Delta G^\circ_{F(TiO)} - \Delta G^\circ_{F(Al_2O_3)}$$
$$= -RT\ln\left\{\frac{a(Al)^2\, a(TiO)^3}{a(Ti)^3\, a(Al_2O_3)}\right\} = -RT\ln\left\{\frac{a(Al)^2}{a(Ti)^3}\right\} \qquad (5.4)$$

In the final simplification, the activities of the solid oxides are taken as unity.

The reaction cited above will progress to the right and TiO will be formed until the activity ratio of the aluminium and titanium reaches an equilibrium value of 1.01 at 1125 °C, or the titanium activity falls below the value (4×10^{-4} at 1125 °C) needed to inhibit the formation of Ti_2O_3 by the degradation reaction

$$3TiO \rightleftarrows [Ti] + Ti_2O_3 \qquad (5.5)$$

At even lower activity levels (below 9×10^{-7}) Ti_3O_5 will be formed, then Ti_4O_7 and ultimately TiO_2 [11].

Which product is formed is important because, while they are all ceramics, TiO differs from the others in being metallic, as indicated by its low electrical resistivity (3×10^{-4} Ω m compared with 0.3 Ω m for TiO_2 and about 10^{12} Ω m for Al_2O_3 [12]). Further, there is evidence that decreasing oxygen-to-titanium ratios can enhance wettability: the contact angles of copper in a vacuum at 1150 °C decrease from 140° on TiO_2 to 113° on Ti_2O_3 to 82° on $TiO_{1.14}$ and to 72° on hypostoichiometric $TiO_{0.86}$ [4].

Table 5.1 Activity coefficient data for some binary alloys [13]

Solvent	*Temperature (K)*	*Activity coefficient**	
		Ti	*Al*
Au	1400	~1	0.00002
Cu	1400	~1	0.008
Ni	1800	0.53	0.00006

*Activity coefficient (activity/atom fraction) as the concentration tends to zero.

These calculations have referred to the activities of titanium and aluminium in solution but have not identified the solvent. In fact the nature of the solvent does not affect the activities needed to form the various titanium oxide reaction products, but it does affect the concentration of titanium needed to form the products. This is because the activity coefficients vary with the nature of the solvent as indicated in Table 5.1.

The variations in the activity coefficients of aluminium are large and influence the minimum titanium concentration needed to form TiO. The low coefficients in gold and nickel mean that the aluminium activity builds up slowly and hence dilute alloys may become depleted in titanium before the equilibrium activity ratio is reached. Thus threshold concentrations for the formation of stoichiometric TiO can be calculated: less than 1 at.% of titanium in copper but more than 4 at.% in gold at 1127 °C and about 10 at.% in nickel at 1527 °C.

These predictions are in good agreement with the experimental results of Naidich and Zhuralev [14] who used X-ray diffraction techniques to observe the presence of TiO at the interface formed by alumina with a Cu–2at.%Ti alloy at 1150 °C, while Ti_2O_3 was found at the interface formed by 2 at.% Ti in gold and nickel at 1500 °C. The beneficial effect of TiO formation at interfaces formed with alumina is suggested by the similarity of the minimum titanium concentrations needed to induce wetting – less than 1 at.% in copper and 3 at.% in gold at 1150 °C and 6 at.% in nickel at 1500 °C. In practice, significantly higher titanium concentrations, typically 8 at.% (6 wt.%) in copper, have to be employed to achieve rapid and reproducible wetting, possibly because of the desirability of forming hypostoichiometric TiO. However, it is not satisfactory merely to promote wetting by progressively increasing titanium concentrations in binary alloys: the alloys become stiff and brittle and can react excessively with their ceramic and metal workpieces.

The ideal braze solvent would be one in which the activity coefficients of the reactive solutes are large. Since saturated solutions have high activities,

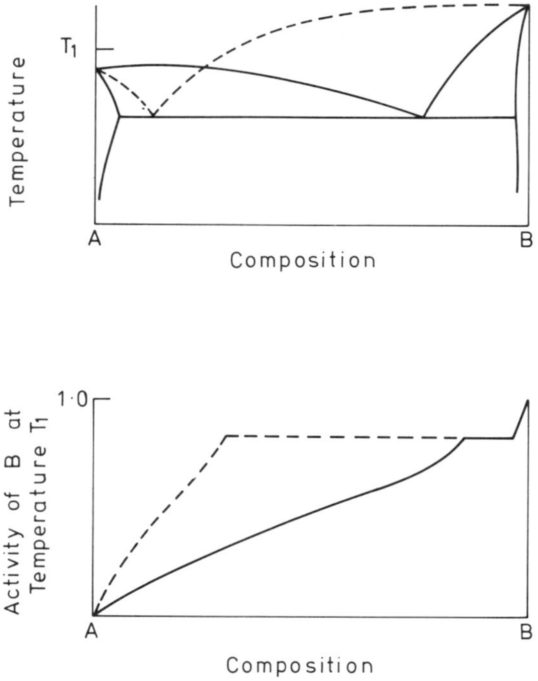

Fig. 5.2 Schematic illustration of the effect of varying solubility on solute activity.

this implies that alloying the solvent to decrease the solubilities can be beneficial, as illustrated schematically in Fig. 5.2. In this figure the variation in the activity of solute B with its concentration at temperature T_1 is illustrated for a system in which B is very soluble in A (full curves) and a system in which B is slightly soluble in A (broken curves). The point of interest is that the activity increases more rapidly with concentration for a low-solubility system, and hence the activity is greater for a given concentration of B.

The solubility of titanium at 1150 °C in various solvents decreases in the order copper (67%), indium (9.4%), tin (7.6%), silver (3.7%) and gold (3.2%) [10]. Practice shows that the addition of silver and gold to copper enhances the ability of titanium solutes to induce wetting, but the effect is even more marked for additions of indium and tin. The reason why indium and tin are more effective than silver or gold as an additive may be due to their efficiency in lowering the surface energy of the alloy. If the titanium solute is active enough to cause wettable TiO to be formed, from the Young equation [2] the extent of wetting will be greater for a liquid with low surface energy.

Figure 5.3 [15] shows that the addition of 25% In decreases the titanium

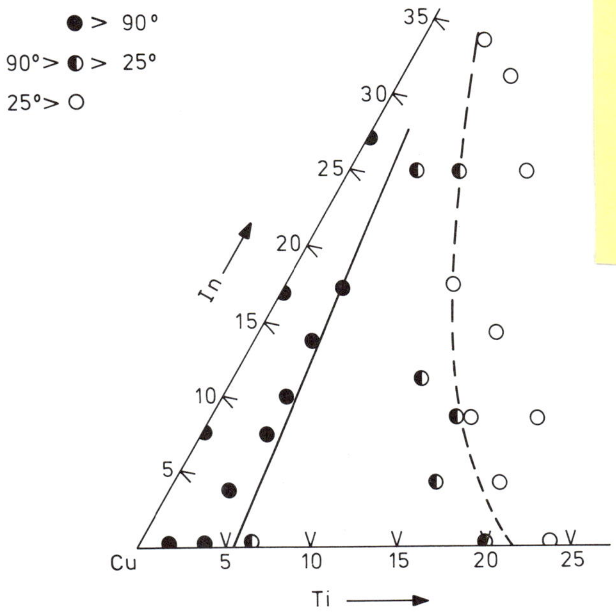

Fig. 5.3 The effect of composition on the wetting of alumina by Cu–In–Ti alloys.

concentration needed to wet alumina reproducibly from 6% to 3%. The decreases in the titanium concentrations needed to cause good wetting, as identified by contact angles of less than 25°, are even more dramatic: additions of 25% In decrease the titanium concentration from 21% to 6%. These data refer to alumina, but similar observations have been made for other oxide ceramics such as silica, zirconia and magnesia. Thus the pattern of behaviour revealed by Fig. 5.3 is probably typical of that for many oxides.

5.2.3 Role of titanium in the wetting of non-oxide ceramics

The wetting behaviour of carbon, carbides and nitrides is not as well characterized as that of oxides, but nevertheless thermodynamic data can be used to gain insight. Thus exposure of carbon to titanium-containing alloys will lead to the formation of TiC. No other product can be formed, but uncertainty can exist about the detailed nature of the reaction because the carbide exhibits a wide range of stoichiometry. The hypostoichiometric compound $TiC_{0.48}$ is stable at 1125 °C when it is in contact with titanium at unit activity, while carbides which are less titanium rich but still hypostoichiometric are stablilized by lower titanium activities [16].

As in the case of oxide reaction products, the chemistry of the interfacial

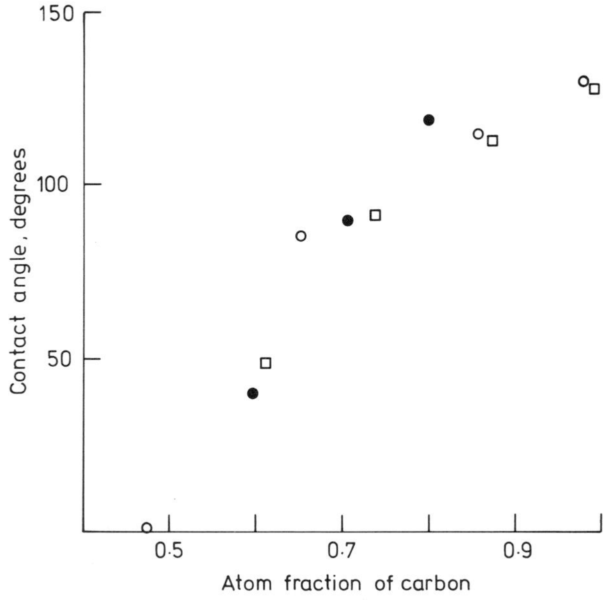

Fig. 5.4 The wettability of titanium carbides by copper at 1150 °C.

carbide is important because of its influence on wettability. Several workers have found bulk TiC to be progressively better wetted as it becomes more hypostoichiometric (Fig. 5.4) [14,17,18], and hence the formation of a markedly hypostoichiometric reaction product at the interface should promote wetting. In practice, wetting of carbon and carbides less chemically stable than TiC occurs rapidly and reproducibly when the concentration of titanium in copper exceeds about 8%, sufficient to stabilize $TiC_{0.65}$ under equilibrium conditions.

While inducing wetting satisfactorily, 8% Ti in copper makes the alloy too reactive to use for some brazing applications such as infiltration of carbon fibres. Once more the titanium concentration needed to promote wetting can be decreased by simultaneously alloying the copper with metals such as gold or silver, and particularly indium or tin. Introducing 9% Sn decreases the titanium concentration needed to wet carbon reliably from 8% to less than 1% [19]. Similar effects have been observed with carbides. Thus Yupko [20] found additions of tin to promote the wetting of SiC by Cu–Ti alloys, while other workers [21] have brazed SiC using an Ag–26Cu–2Ti braze.

Detailed reactivity data for nitrides are even more sparse, but this lack may be remedied soon because of the interest in these ceramics as engineering and electronic materials. Their interactions with titanium-containing alloys are probably similar to those of carbides because, once again, there is only one high-temperature reaction product – titanium mononitride – which has a stoichiometry range extending to $TiN_{0.43}$ at 1150 °C [10]. Bonding techniques for nitrides are still being developed, but the technical literature contains references to the use of an amorphous Cu–Ti alloy to metallize Si_3N_4 [22] and of Ag–27Cu–2Ti, Ti–28Ni and Ti–49Cu–2Be to braze Si_3N_4. There are also references to the use of pure aluminium and Al–Si alloys [9].

5.2.4 Role of titanium – application to commercial brazes

Having described some of the fundamental data and phenomena underlying the behaviour of brazes, we can now reconsider the compositions of commercial braze alloys. One of the more notable features of the use of commercial brazes is that the same compositions are employed to join a wide range of ceramics. For example, alumina and other oxides, carbon, carbides and nitrides have all been brazed with Ag–27Cu–2Ti and Ag–23Cu–15In–1.5Ti.

Titanium is the activator used in commercial brazes, reflecting the singular characteristic that it can form very wettable reaction products. Other metals, such as hafnium, zirconium, tantalum or chromium, can also form potentially wettable reaction products, but their use as activators does not confer advantages except for a few specific applications. Thus although alternative alloys can be purchased, they are special products that usually command a premium. At present, therefore, titanium-activated brazes dominate the commercial market.

The general use of an Ag–Cu alloy rather than pure copper as a solvent is understandable in terms of effects on the solubility and activity of titanium and hence the ease of forming a wettable reaction product. Similarly, the introduction of indium and tin as quaternary alloying elements is consistent with their greater enhancement of titanium activity, and recent work has identified TiO as the reaction product formed by the Ag–26Cu–4Sn–4Ti and Ag–23Cu–15In–1.5Ti alloys with ZrO_2 [6].

Ag–Cu solvents also offer other advantages. The preferred composition Ag–28Cu is a eutectic, and hence has a singular melting temperature of 780 °C. The solvent alloy is ductile, enabling preforms to be fabricated readily and conferring on joints some ability to accommodate thermal mismatch stresses. The titanium additions can affect the alloy melting points, causing the liquidus and solidus to separate so that melting involves a transitional two-phase stage and stiffening and hardening the solidified alloy. The introduction of tin and indium will also stiffen and harden the solvents, but the Ag–23Cu–15In–1.5Ti alloy is described as being ductile, so that these complex alloys can still have attractive or at least acceptable properties.

5.2.5 Other titanium-induced brazing processes

The brittleness and difficulties of fabricating many active metal brazes has prompted interest in ways of promoting beneficial titanium-induced reactions other than by alloying. Reference has already been made to the use of unalloyed titanium as an inner layer in composite braze sheet or core in composite wire. Similarly, ceramic surfaces have been vapour or ion coated with metallic titanium prior to brazing. Of direct relevance to this chapter is the use of unstable TiH_2 coatings and the practice of using pure titanium, or zirconium, as the braze material.

The TiH_2 process consists of applying the compound to the ceramic surfaces as fine mesh material suspended in fugitive carrier or binding agents such as aqueous ethyl glycol, methyl methacrylate or commercial braze cements. The hydride dissociates when heated in a vacuum to 300–500 °C to leave a layer of fresh metallic titanium on the ceramic surface which can react to adhere. Thus the process need merely involve a dwell or slow ramp through the dissociation range as the workpieces are heated to the brazing temperature. Conventional titanium-free braze alloys will wet the ceramics where the TiH_2 has been applied successfully but not elsewhere, and hence their flow is controllable.

Many of the details of this technique are not characterized or described in the open literature, but considerable success has been achieved by some fabricators who exercise careful control and employ application methods based on skilful manipulation. Thus particular care has to be taken in the selection of the fugitive carrier or binder – the fluidity, uniformity and storage behaviour of the suspension are all of crucial importance – and the application of a uniformly thick layer in the desired regions is frequently a manual skill akin to those employed in tableware decoration.

In contrast, the use of titanium or zirconium foils as preplaced braze, usually for face-seal joints, is a technological development based on the fact that these reactive metals form eutectics with iron, nickel, copper and a range of alloys that find use as common workpiece materials [10]. In applying the technique decisions have to be taken not only about the fabrication temperature and time but also about the thickness of foil to be used, taking into account the conflicting need to fill the gap between the workpieces with a melt while causing as little erosion of the metal workpiece as possible.

5.3 BRAZING

Many factors must be taken into consideration in choosing the fabrication conditions to be employed to braze a ceramic component. Three of the most important, the selection of surface cleaning and preparation processes, the environment and the thermal cycle, are discussed in this section.

5.3.1 Surface Preparation

Surface preparation methods for metal components prior to their brazing to ceramics are well established. There is clear evidence that surface roughness affects wetting and flow behaviour during brazing, and the roughness of the metal surface can be varied by machining or abrasive operations. In practice, surface roughness is generally determined by the component manufacturing process – milling, drawing and so on – but mechanical preparation to remove blemishes such as score marks can be beneficial. It is seldom desirable to grit-blast or necessary to produce a mirror finish.

Chemical treatments or abrasion are used to ensure that metal surfaces are not coated by thick oxide films – even though thin films will reform virtually instantaneously. The surfaces are then treated with degreasing agents to remove organic contaminants that could degrade to form carbonaceous layers during brazing. The desire to remove thick oxide films and avoid carbonaceous layers may be thought surprising since the braze alloys to be used can wet oxides and carbon. However, the surface films and their reaction products may not be firmly attached to their metal substrates and hence the joints may be weakened.

The roughness of ceramic surfaces can also affect wetting and flow behaviour, with the rougher surfaces degrading wettability except by the most efficient braze alloys, i.e. those with contact angles less than 20° [23]. However, ceramic component surfaces generally have to be ground to meet engineering tolerances of size and shape, and grinding can not only produce a surface roughness but can also nucleate cracks which degrade mechanical properties. Thus for strong joints it may be necessary to finish the ceramic surfaces by lapping or refiring at a high temperature to blunt or heal the cracks. Such firing, if in air, can also remove loosely adherent carbonaceous films from oxide surfaces. However, removal of oxide films from the surfaces of nitrides or carbides is best done mechanically and followed by a degreasing process.

5.3.2 Environment

The environment in which active metal brazing is conducted is of considerable importance because of the necessary reactivity of the titanium. In most operations a vacuum is employed to achieve low oxygen and nitrogen activities, with furnace pressures of less than 10^{-3} mbar being recommended and vacua of 10^{-4}–10^{-5} mbar being employed when possible. However, it should be borne in mind that some dissociation of nitrides can occur in environments with low nitrogen activity.

The threshold pressure at which brazing becomes unreliable depends on several factors, including the braze ratio of workpiece surface area to furnace volume and most importantly the leak tightness of the furnace chamber. Even if a leaky chamber is evacuated to a low pressure, the residual gas is mainly air and hence the oxygen and nitrogen activities approach 0.2 and 0.8 of the total pressure expressed in atmospheres. In a leak-tight system the residual gas derives mainly from outgassing by the chamber walls and furniture, and typically the oxygen and nitrogen activities are 10–100 times less than in a leaky system. Therefore in determining the conditions to be used and then controlling them, it is useful to monitor not only the vacuum level but also the partial pressures of the residual gases.

If vacuum brazing is impractical, the less common practice of brazing in reducing or inert gas atmospheres may be possible. However, considerable care must be taken to ensure the quality of the gaseous environment employed. While it is relatively easy to achieve an oxygen activity of 10^{-6} by evacuating a leak-tight chamber to about 5×10^{-3} mbar, argon or helium with an oxygen content of 1 ppm requires careful handling if it is to remain pure. Often some repurification process is required, such as passage over titanium swarf at about 900 °C immediately before entry into the brazing furnace. Similarly, if a reducing hydrogen environment is used, the dew point should be lower than −51 °C.

5.3.3 Thermal cycle

The thermal cycle can be closely controlled by the fabricator and hence its selection is particularly important. The specific heating rates employed will depend upon the capabilities of the equipment and the size and geometry of the workpieces. Radio frequency and induction heating rates are potentially far more rapid than those achievable in resistance furnaces but precise temperature control is more difficult, and in practice heating and cooling rates may be restricted by the need to avoid thermal shock to large components or those with very variable cross-sections.

A common and often successful practice is to heat the workpieces moderately slowly, perhaps at 10–20 K min^{-1}, to a temperature just below the braze solidus and to hold for at least 15 min to ensure uniformity of temperature within the workpieces. (If TiH_2-activated brazing is being used, a hold at 350–500 °C to allow dissociation is also necessary.) This subsolidus hold is then followed by even slower heating to the brazing temperature, which is often 50 °C or more above the liquidus, and a hold for about 30 min. The slow heating (1–5 K min^{-1}) helps to maintain temperature uniformity throughout the component and to avoid premature flow of the more readily melted and probably inactive component of the alloy over the wettable metal components – sometimes called 'blushing' – to leave a high-melting-point residue in the capillary gap formed with the ceramic. The long dwell time at the brazing temperatures relates to the usually poor fluidity of the brazes or the need for them to react with the workpieces if a wetting front is to advance or to both.

It is more difficult to give guidance about optimum cooling rates, which can be varied by adjusting power levels and backfilling with inert gases, because they will be controlled by a balance between the need for a high throughput on the one hand and mitigation of unacceptable thermal shock and high thermal contraction mismatch stresses on the other. Therefore the optimum rate is a function more of the nature of the component, its size and the properties of the workpieces than of the braze. Consideration needs to be given to the design of joints and the stresses within them in making an initial selection. In view of the frequent complexities of the thermal cycles and the importance of close control it is recommended that programmable power input devices are used whenever they are available as an aid to reproducible production and troubleshooting.

5.4 BRAZEMENTS

The range of properties required of brazed ceramic joints can be very varied, but most can be regarded as aspects of physical, chemical or mechanical integrity. Among the most commonly cited properties are leak tightness, environmental stability, thermal endurance and strength.

Physical integrity as measured by leak-tight joints between metals and ceramics is produced routinely by compression in vacuum systems using, for example, gold wire as a gasket material to seal silica windows to steel flanges. Therefore, for a brazed joint to be leak-tight it should merely be necessary that the braze metal and the ceramic be in contact, i.e. wetting has produced a continuous physical barrier of braze metal in the capillary gap. The gap does not have to be filled completely, and it is sufficient for the metal–ceramic interface to exist rather than be bonded strongly. Of course in practice leak-tightness is usually a result of a joint being strongly bonded, but leak-tightness as such is not an indicator of good bonding.

Physical integrity can be measured by simple techniques, but assessment of the chemical integrity of a joint is more complex. If this might be degraded by the service environment, care must be taken to use or simulate those conditions and take account of transient environmental variations. Interdiffusion in service conditions may also be of concern since this could lead to the growth of fragile reaction product layers at the interfaces formed with either the metal or the ceramic workpieces.

Thus mechanical integrity, the commonest measure of joint quality, may be degraded by service-induced chemical changes. It is also affected by the chemical reactions necessary to ensure the wetting of ceramics by active metal brazes. Some of these effects are discussed below.

5.4.1 Metal–ceramic strengths

Defining an acceptable measure of the mechanical integrity of a joint is difficult. Selection of an appropriate test technique can be complicated because, while national and international procedures have been established for monolithic materials, those for ceramic–metal interfaces are still being developed.

The ASTM method for measuring the tensile strength of brazed joints between metallized ceramics [25] has been adopted by some workers for assessing the strengths of active metal brazed ceramics. However, many reported tensile strength values have been derived from other experimentally convenient but non-standardized methods such as measurement of the push-off loads for solidified sessile drops and the bursting pressures of capped tubes (Fig. 5.5). Partially in response to the need for design data, fracture mechanics techniques are now being developed as described in Chapter 8. Fortunately, even though the absolute strength values suggested by the ASTM and non-standardized tests can differ considerably,

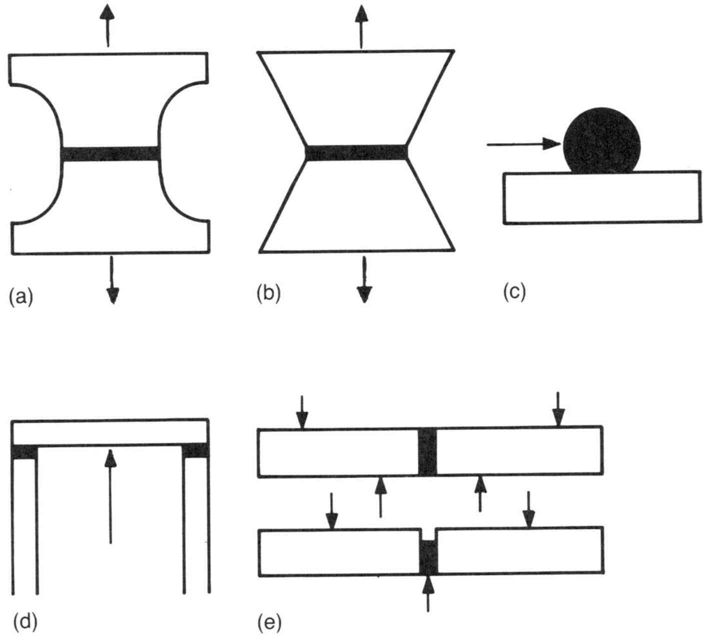

Fig. 5.5 Schematic illustration of some tests used to assess the strengths of ceramic–metal interfaces: (a) ASTM tensile test; (b) truncated cone tensile test; (c) sessile drop push-off test; (d) capped-tube blow-off test; (e) notched and unnotched bend tests.

the conclusions derived from the tests about the influence of fabrication and material parameters are generally consistent.

Comparatively few data are available that characterize the strengths of ceramic joints brazed with pure metals because these systems are generally non-wetting and hence brazing is impractical. Nevertheless, push-off load data derived from solidified non-wetting sessile drops indicate that bonding to ceramic substrates can be strong: tensile strengths for non-wetting aluminium–alumina interfaces are estimated as 90 MPa [26]. For chemically unreactive systems, the tensile strengths derived from this test are not appreciably affected by changes in the extent of non-wettability induced by varying the contact temperature or time, suggesting that strengths reflect material rather than fabrication parameters.

Models of ceramic–metal interfaces generally relate only to material characteristics, and McDonald and Eberhart [27] suggested a correlation between the thermodynamic work of adhesion of molten metal–alumina interfaces ($W_{AD} = \gamma_c + \gamma_m - \gamma_{cm} = \gamma_m(1 + \cos\theta)$) and their room temperature strengths. This correlation is poor, but strengths can be compared with work of adhesion values for a few solid metal–alumina interfaces for which data are available (Fig. 5.6) [26,28]. The metals forming these interfaces should have been oxide free in the fabrication conditions employed, and hence according to the McDonald–Eberhart model both the work of adhesion and strength data should have been influenced by the affinity of the metal atoms for the oxygen ions dominating the alumina surfaces. It is therefore of interest, and possible value, that the strengths plotted in Fig. 5.6 are also ranked correctly by the ΔG°_{oxide} and melting-point data of the metals. Whether such rankings are also applicable to non-oxide ceramics is not yet clear.

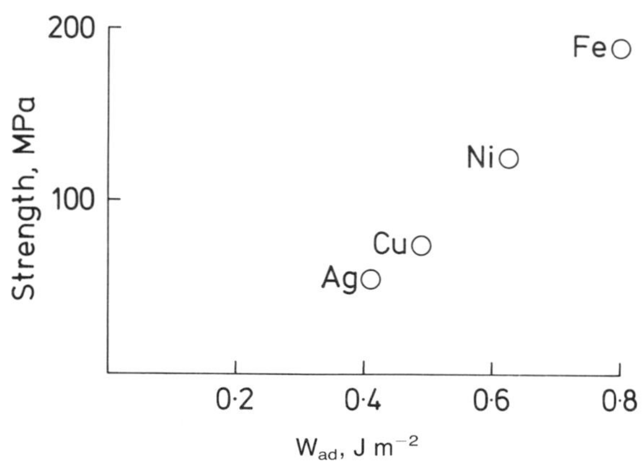

Fig. 5.6 Room temperature tensile strengths for pure metal–alumina interfaces compared with solid–solid work of adhesion values.

5.4.2 Binary alloy–ceramic strengths

Pure metals are seldom used to braze ceramics, and the wetting ability of active metal braze alloys depends on the occurrence of chemical reactions at the molten metal–ceramic interfaces. However, these reactions are preserved in the chemistry and structure of the solidified metal–ceramic interfaces and hence can affect strengths. The reaction products can be regarded as chemical bridges between the two materials and studies with transparent ceramics – such as sapphire and diamond [29] – show that the bridges are nucleated as islands that grow in number and size as the reaction proceeds. The progressive increase in the total area of the bridges with increasing concentration, or rather activity, of the reactant, brazing time and temperature is accompanied by an increase in the strength of the interface. Ultimately, the interface is completely covered by the bridging layer and the strength reaches a maximum value. Further reaction merely thickens the bridging layer and generates volume mismatch and other stresses which become progressively more difficult to accommodate. The layer becomes flawed and may even separate from its substrate, as rust does from steel. As a result, the strengths of the interfaces decrease. Thus there is an optimum extent of reaction which occurs when the chemical bridges first coalesce to form a complete layer. In the case of Cu–Ti alloys bonded to diamond by contact at 1150 °C for 0.3 h, the optimum extent of reaction is that which produces a complete layer with a thickness of 0.17 μm [29]. Exceeding this optimum can have a dramatic effect on the strength of the interface (Fig. 5.7).

Strength maxima have been observed for many systems. Although difficult to define with precision, their apparent values for bonded oxides

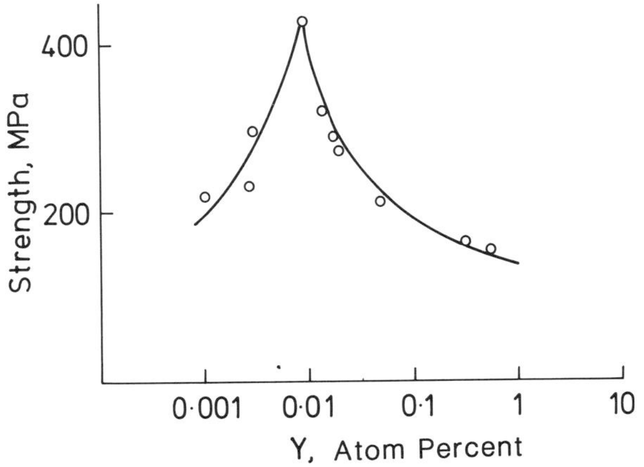

Fig. 5.7 Room temperature tensile strengths for interfaces formed by Ni–Y alloys with alumina.

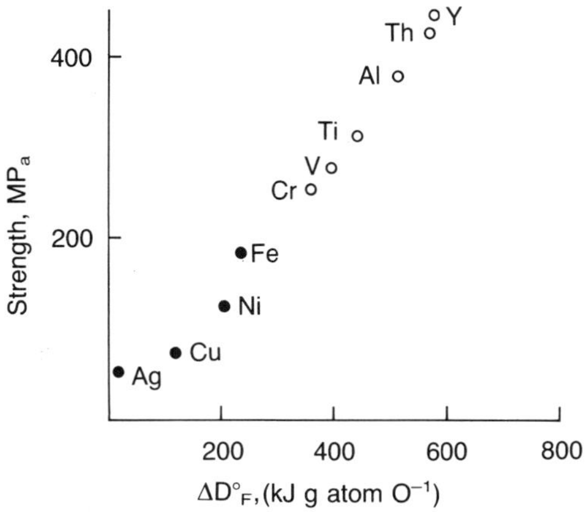

Fig. 5.8 Comparison of solute oxygen affinity and the maximum tensile strengths of interfaces formed by alumina with various nickel binary alloys (○) and pure metals (•).

can be related to the chemical affinity of the active component of the alloy for oxygen (Fig. 5.8) [30] in good accord with the McDonald–Eberhart model. (Included in the figure are the strength data for pure metal–alumina systems previously plotted against the work of adhesion values in Fig. 5.7.) Such trends also occur for alloys bonded to carbon [30], but whether this is also the case for nitride ceramics is not yet clear.

Although of value in gaining understanding of the factors that control the strength of a braze–ceramic interface, the examples cited so far are of no practical interest because the alloy concentrations, contact times and temperature that resulted in achieving the optimum extent of reaction and bonding did not also result in wetting. Thus the fabricator will often be faced with the problem of wetting requiring the formation of excessively thick and fragile reaction product layers. What is needed is an alloy in which the reactant activity is high but the rate of reaction product formation is low, so that a wettable reaction product is formed but only as a thin layer. This requirement is similar to that which led to an interest in ternary alloy additives as wetting promoters – a high activity but low concentration – and not surprisingly such additives also affect interfacial strengths.

5.4.3 Ternary and complex alloy–ceramic strengths

The bonding behaviour of ternary and complex alloy systems broadly resembles that described in the previous section. Increasing the content of the reactive metal (titanium) at first increases the interfacial strength, but ultimately weakening occurs so that the systems display bonding maxima and optimum extents of reaction. Varying the contact time and temperature can cause interfacial strength values to peak, as can variations in the concentration of non-reactive ternary and higher-order alloying elements.

Exploitable braze alloy compositions wet the ceramics and strength data can be derived using generally accepted or analysable methods such as ASTM tensile tests [25] or bend and fracture mechanics tests. Thus the effects of braze alloy composition variations and other variables can be more certainly defined and characterized in terms of the behaviour of joints, although the relatively large size of some joint samples may give rise to conservative strength estimates. More guidance can be derived from other chapters but it is clear that braze alloy and fabrication parameter optimization can lead to significant maximization of joint strengths.

The strengths of alumina ASTM test pieces brazed with Cu–In–Ti alloys are a function of both the indium and titanium contents of the alloys as illustrated in Fig. 5.9 [31]. Indium can both enhance and degrade bond strengths, but the data identify compositions of the ternary alloy centred on Cu–10In–30Ti for which bonding is maximized (Fig. 5.9). Similar strength variations have been observed for Cu–Ti–Au, Cu–Ti–In and Cu–Ti–Sn alloys, and these optimized compositions were used to identify the patented ternary brazes cited previously [7]. The peak strength for the

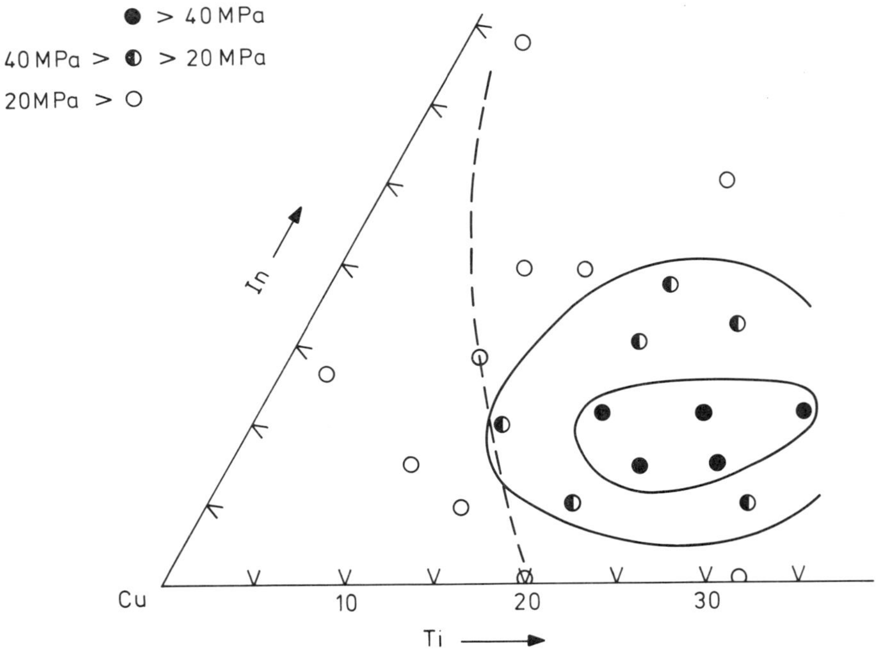

Fig. 5.9 The strengths of ASTM test pieces brazed with Cu–In–Ti alloys.

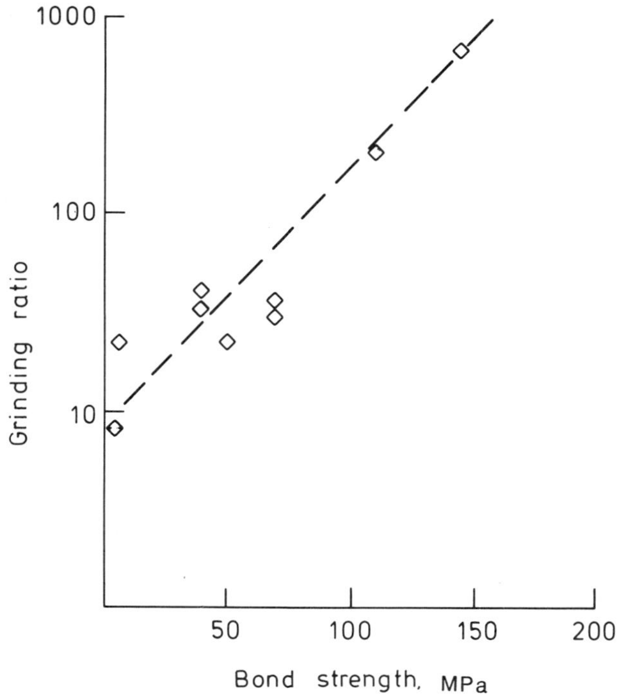

Fig. 5.10 The effect of (Cu–Sn–Ti)–diamond bond strengths on the ability of abrasive compacts to grind steel. The grinding ratio is the ratio of the weight loss of steel to the weight loss of the compact.

Cu–Sn–Ti alloys bonded to alumina was 52 MPa, and this was achieved with a Cu–15Sn–15Ti alloy. This composition also produced the greatest strength (150 MPa) when used to bond diamond [32], as measured in a simple pull-off test, and the approach to this optimum composition had a dramatic effect on performance when Cu–Sn–Ti alloys were used to make abrasive diamond compacts by infiltration (Fig. 5.10).

Other recent examples of the effect of composition or reaction optimization on strength data are provided by the reports of strong shear bonding of zirconia to Ti–6Al–4V, AISI 446 and nodular cast iron using the commercial Ag–23Cu–15In–1.5Ti braze [33] and of strong tensile bonding of alumina using an Ag–35Cu–2Ti braze [34].

It is important to realize, however, that mechanical property data for test samples give only partial guidance to the behaviour of components made by brazing ceramic–metal or ceramic–ceramic workpieces. Differing sizes and particularly differing thermal contraction characteristics of the workpieces can affect mechanical integrity. A large component is more likely to be weak because there is a greater statistical probability that the workpieces or interfaces will contain damaging flaws, and if the workpieces have differing contraction characteristics there is also an increase in the likelihood that such flaws will be generated as the brazed component cools from the fabrication temperature. Thus the tensile strength of laboratory components comprising a sheet of AISI 321 0.5 mm thick bonded to alumina ASTM test pieces of outside diameter 16 mm and inside diameter 10 mm by brazing with Cu–15Sn–15Ti for 0.3 h at 950 °C was 31 MPa, while the tensile strength of components of outside diameter 83 mm and inside diameter 70 mm was only 1 MPa, and similar Ag–27Cu–5Ti brazed joints failed at stresses of about 0.5 MPa [35]. To overcome or accommodate such effects, great care has to be taken over the design of joints.

5.5 CONCLUSION

Active metal brazing can be used to produce high-integrity ceramic–ceramic and ceramic–metal joints. However, wetting and bonding are not synonymous, and hence considerable care is needed in the selection of the brazing alloy and fabrication conditions. In making the selection, caution should be exercised in comparing wetting or strength data derived from work employing different joint configurations or component sizes.

REFERENCES

1. Roberts, P.M. (1987) *Met. Constr.*, **19**, 12.
2. Young, T. (1805) *Trans. R. Philos. Soc.*, **95**, 65.
3. Pauling, L. (1939) *The Nature of the Chemical Bond*, Cornell University Press, Ithaca, NY.
4. Naidich, Yu. (1981) *Prog. Surf. Membrane Sci.*, **14**, 353.
5. Moorhead, A.J., Biterskowitz, M., Morgan, C.S., Woodhouse, J.J. and Read, R.W. (1980) *Fabrication of Sensors for High Temperature Steam Instrumentation Systems*, Rep. ORNL/NUREG–65, Oak Ridge National Laboratory.
6. Moorhead, A.J., Santella, M.L. and Henson, H.M. (1987) *Am. Ceram. Soc. Symp. on Structural Ceramics Joining*, in press.
7. Nicholas, M.G. and Valentine, T.M. (1984) *Ternary alloy brazes*, UK Patent 2066291.
8. Suganuma, K., Okamoto, T.M. Koizumi and Shimada, M. (1987) *J. Mater. Sci.*, **22**, 1359.
9. Kohno, A., Yamada, T. and Yokoi, K. (1985) *J. Jpn. Inst. Met.*, **49**, 876.
10. Massalski, T.B. (1986) *Binary Alloy Phase Diagrams*, ASM, Metals Park, OH.
11. Chase, M.W., Curnutt, J.L., McDonald, R.A., Syverud, A.V. (1975) *J. Phys. Chem. Ref. Data*, **4**, 1. Chase, M.W., Curnutt, J.L., McDonald, R.A., Syverud, A.V. (1978) *J. Phys. Chem. Ref. Data*, **7**, 793.
12. Samsonov, G.V. (1982) *Oxide Handbook*, Plenum, New York.
13. Hultgren, G.V. (1973) *Selected Values of Thermodynamic Properties of Binary Alloys*, ASM, Metals Park, OH.
14. Naidich, Yu. and Zhuralev, V.S. (1974) *Refractories (USSR)*, **15** (1–2), 55.

15. Nicholas, M.G., Waite, M.J. and Valentine, T.M. (1980) *J. Mater. Sci.*, **15**, 2197.
16. Storms, E.K. (1967) *Refractory Carbides*, Academic Press, New York.
17. Ramqvist, L. (1965) *Int. J. Powder Metall.*, **1**, 2.
18. Mortimer, D.A. (1975) *The Effects of Composition on Some Properties of Titanium Oxycarbide*, UKAEA Rep. AERE-R7942, Atomic Energy Research Establishment, Harwell.
19. Standing, R.J. and Nicholas, M.G. (1978) *J. Mater. Sci.*, **13**, 1509.
20. Yupko, V.L. (1977) *Sov. Powder Metall. Met. Ceram.*, **16**, 718.
21. Iseki, T., Matsuzaki, H. and Boadi, J.K. (1985) *Bull. Am. Ceram. Soc.*, **64**, 322.
22. Naka, M., Kubu, M. and Okamoto, I. (1986) *J. Mater. Sci. Lett.*, **5**, 855.
23. Hitchock, S.J., Carroll, N.T. and Nicholas, M.G. (1981) *J. Mater. Sci.*, **16**, 714.
24. Lugscheider, E., Krappitz, H. and Mizuhara, H. (1985) *Fortschr. Dtsch. Keram. Ges.*, **1**, 199.
25. *Standard Method for Tension and Vacuum Testing Metallized Ceramic Seals*, ASTM F19–64, 1964.
26. Nicholas, M.G. (1968) *J. Mater. Sci.*, **3**, 571.
27. McDonald, J.E. and Eberhart, J.G. (1965) *Trans. AIME*, **233**, 512.
28. Pilliar, R.M. and Nutting, J. (1967) *Philos. Mag.*, **16**, 181.
29. Scott, P.M., Nicholas, M.G. and Dewar, B. (1975) *J. Mater. Sci.*, **10**, 1833.
30. Crispin, R.M. and Nicholas, M.G. (1976) *J. Mater. Sci.*, **11**, 17.
31. Nicholas, M.G. (1981) *DVS Ber.* **66**, 29.
32. Evens, D., Nicholas, M.G. and Scott, P.M. (1977) *Ind. Diamond Rev.*, **12**, 306.
33. Moorhead, A.J. and Santella, M.L. (1987) *Am. Ceram. Soc. Symp. on Structural Ceramics Joining*, in press.
34. Mizuhara, H. and Huebel, G. (1987) *Am. Ceram. Soc. Symp. on Structural Ceramics Joining*, in press.
35. van Esdonk, J. and van der Sluis, H.H. (1983) *Brazing and Soldering*, **5**, 22.

6

Diffusion bonding

B. Derby

6.1 INTRODUCTION

Diffusion bonding is a joining process in which two nominally flat surfaces are held together at an elevated temperature (typically above 60% of the melting point of the least refractory material) for a period of time until a bond is formed. This process is relatively simple when two identical materials are to be joined, but when joining dissimilar materials there are many potential complications. In order to limit discussion, in this chapter we shall consider chiefly solid state diffusion bonding where all phases present remain solid. Thus we shall not describe or discuss transient liquid phase joining where an interfacial reaction forms a liquid phase which later solidifies by a diffusional equilibration. However, this definition does include diffusion bonding, reaction bonding, direct bonding and field-assisted bonding. These earlier definitions of metal–ceramic joining techniques all promote bonding by similar mechanisms at the interface.

Diffusion bonding of metals has a long history. It was one of the first joining technologies developed by man, and examples of artefacts fabricated in this manner have been discovered in very early sites of metal working [1]. The application of diffusion bonding metals to ceramics has a more modest history, growing out of early work on glass–metal seals. In principle the technique can also be used to join ceramics to ceramics. Indeed this has been successfully done in a number of cases [2,3]. However, the need for temperatures of above 50% of the melting point of the materials being joined limits its practicality for most high-temperature ceramics.

Diffusion bonding is a solid state process and any deformation of the joining components occurs only in the near-surface region. It is a near net shape joining technique which is particularly suited to applications requiring high degrees of accuracy and has a number of advantages over brazing. All constituents in the process are solid, and thus the thermodynamic driving force for interface formation is simply the work of adhesion

$$W_{ad} = \gamma_{S_1V} + \gamma_{S_2V} - \gamma_{S_1S_2} \quad (6.1)$$

where g is the energy of the interface between the two solid phases S_1 and S_2 and V (vacuum) denotes a free surface. In brazing the liquid metal must wet the ceramic (see Chapter 5) and the driving force for wetting is $W_{ad} - \gamma_{LV}$ where γ_{LV} is the liquid braze metal surface energy and W_{ad} is the work of adhesion for a solid–liquid interface. Hence a low work of adhesion may result in the formation of a good joint by diffusion bonding but not by brazing. In addition, the diffusion bonding method allows the simultaneous manufacture of many bonds in a complex joint configuration by stacking thin strips of material and subjecting the assembly to bonding conditions. This approach allows some mitigation of the thermal expansion problem in metal–ceramic joining and will be dealt with in more detail later. A final potential advantage of diffusion bonding over brazing is that alloys with comparatively low melting point do not have to be used as fillers for ease of fabrication, and hence the inherent upper temperature for the use of diffusion-bonded joints in service is high.

Diffusion bonding is therefore a potential contender for applications in high-temperature environments, e.g. in adiabatic engine components and ceramic turbo-chargers. A number of applications of the engineering use of metal–ceramic joints have been reported where diffusion bonding has been used successfully. One such example is in the accelerator modules of the Daresbury nuclear structure facility described in Chapter 11. Another application is in the fixing of dielectric windows of quartz and sapphire onto the large toroidal steel vacuum chamber of the JET fusion experiment at the Culham Laboratory [4]. A final example is the use of diffusion bonding to seal the Na–S battery electrolyte [5].

The details of the bonding procedure will vary from case to case but common points exist. Diffusion bonding occurs between contacting surfaces; hence before attempting to bond the surfaces must conform. This is usually achieved by grinding the surfaces to nominal flatness. The surface condition is an important process variable and, in common with all joining techniques, cleanliness is very important. It is known that the degree of surface roughness is another important parameter in diffusion bonding, ranking in importance with temperature and pressure in determining bonding rates [6].

In this chapter we shall present details of diffusion bonding mechanisms and how they depend on process conditions. (The use of diffusion bonding in a technological application is described in Chapter 11.) The possibility of chemical reactions and the influence of the ambient conditions of such reactions are discussed. Finally, the strength of successful bonds will be reviewed for a range of metal–ceramic combinations.

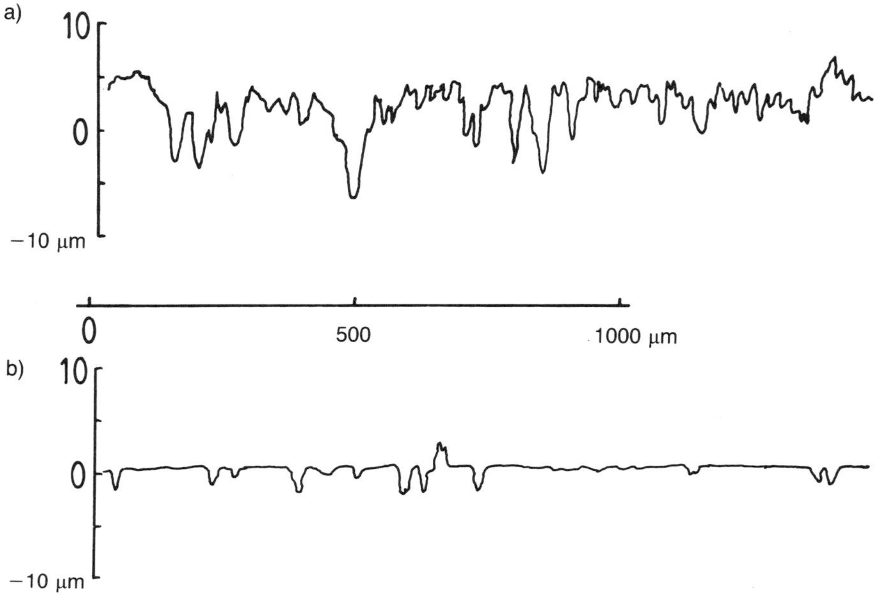

Fig. 6.1 Two alumina surfaces: (a) after sintering; (b) after diamond grinding with 3 μm particles.

6.2 MECHANISMS

We have defined diffusion bonding as the creation of an intimate bond or joint between two materials by thermally assisted processes occurring in the solid state. In order to understand the mechanisms and driving forces of diffusion bonding, the evolution of the bond microstructure must be appreciated. Prior to bonding, two nominally flat surfaces exist in overall contact, but the actual extent of contact will depend critically on the roughnesses of these surfaces. Some typical surface roughnesses are shown in Fig. 6.1. Clearly any surface finish will depend on the final grinding or lapping process used. In practice we need to know how surface roughness influences bonding mechanisms so that a minimum of surface conditioning can be used. After diffusion bonding has occurred, the result should be a strong interphase boundary. Thus the formation of a bond must allow the transition from two contacting surfaces to an intimate joint (Fig. 6.2).

The bonding process can be broken down into two necessary steps operating in parallel. The first is the transition from two surfaces contacting at their asperities to an intimate interfacial conformity, as illustrated in Fig. 6.2. This must involve the elimination of a large volume of interfacial voids accommodated by mass transfer mechanisms (e.g. plastic flow and

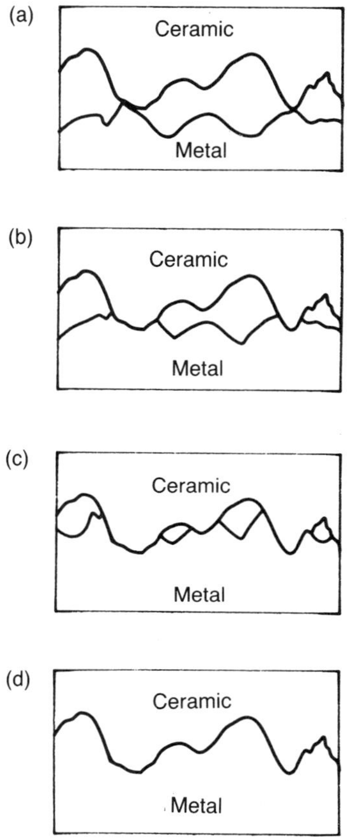

Fig. 6.2 Sequence of events during metal–ceramic diffusion bonding: (a) hard ceramic and comparatively ductile metal surfaces come into contact at asperity tips; (b) ceramic asperities indent the metal surface which yields plastically under the large local stress; (c) deformation continues primarily in the metal, coupled with diffusional mass transfer leading to a shrinkage of the interfacial voids; (d) finally the bond is formed, ideally as a perfect interphase boundary.

diffusion). In parallel with this, but sequential to each individual contact, there must also be an adhesion process giving the interphase boundary strength. A third step, with possible deleterious consequences, is a subsequent chemical reaction between the metal or ceramic in contact to form a third phase at the boundary.

The diffusion-bonding process between two identical metal surfaces is known to be determined by the mass transfer processes which lead to void collapse and an intimate contact at the joint [7,8]. The transformation from good contact to a strong bond presumably takes place by relaxation of the atom locations from those characteristic of two surfaces to those of a grain boundary, and this process is very rapid. When two dissimilar materials are brought into contact the determining process is not clearly definable. The rate of void collapse is certainly important, but the thermodynamics and chemistry of the interface are now much more complicated. The initial ceramic–metal contact will not be in equilibrium and the processes which achieve equilibrium may take some time. These complexities are evident in one observation of ceramic–metal diffusion bonding which is very different from metal–metal bonds, namely the presence of a critical temperature below which no bond is observed. This has been reported in a number of systems with various critical temperatures, e.g. $Al-Al_2O_3$ (450 °C), $Ag-Al_2O_3$ (600 °C), $Cu-Al_2O_3$ (700 °C) and $Ni-Al_2O_3$ (900 °C) [9] and also $Pd-ZrO_2$ (1100 °C) [10]. Clearly, the adhesion reaction is important in ceramic–metal joining and needs to be addressed if a full understanding of diffusion bonding is to be achieved.

6.2.1 Interface Formation

In this section we describe the evolution of an intimate interfacial contact during diffusion bonding; the complex interfacial interactions which are the cause of a strong ceramic–metal interphase boundary are treated in Chapters 5 and 7. The mechanisms of diffusion bonding two identical surfaces have been studied since the 1960s, and there is now general agreement that joint formation occurs by collapse of interface voids produced by a number of diffusional and creep mechanisms. Several papers on this topic have been published in recent years [6,11–13] and developments have been reviewed [14]. The collapse of interfacial voids can be brought about by a number of mechanisms [6], and these are best grouped in terms of sources and sinks for matter as illustrated in Fig. 6.3 and Table 6.1.

Bonds made between dissimilar materials will have a more complicated combination of bonding mechanisms [15]. The same void-closure mechanisms will operate but many more diffusion pathways may occur. In addition, the relative rates of component diffusion, coupled with multi-component interfacial chemistry, can lead to kinetically favoured non-equilibrium structures. The kinetics of dissolution and material transport can result in interface structures dependent not only on bonding conditions but on the rate of cooling from the bonding temperature [16]. At this point we must carefully distinguish diffusion bonding mechanisms from other local mass transport effects.

Interface formation must be accompanied by the collapse and annihilation of voids created at first contact. The driving force for this collapse is identical with that which drives diffusion bonding in metals. Void closure results in a net approach of the two surfaces being joined. This allows mechanical work to be done by the bonding pressure. The reduction in

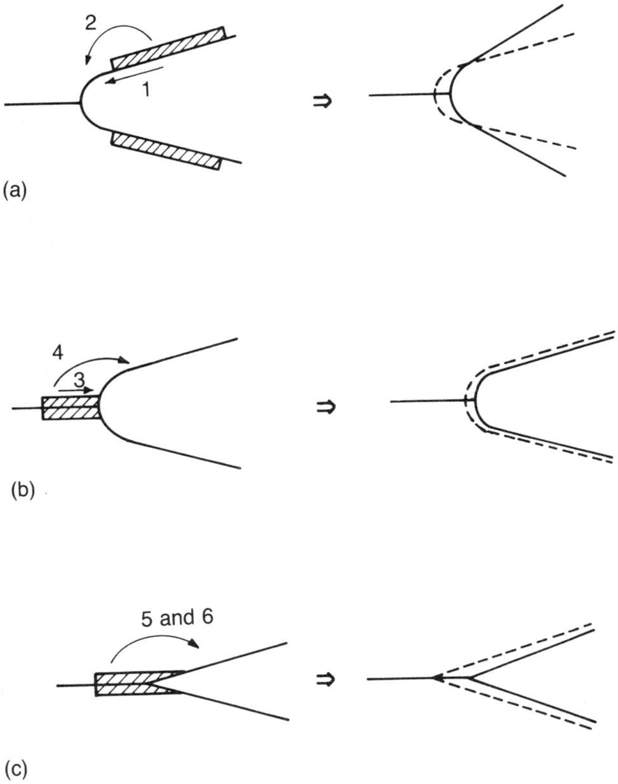

Fig. 6.3 Sources and sinks of mass transfer during diffusion bonding (see also Table 6.1); (a) transport from surface sources to an interfacial neck; (b) transport from the bonding interface to a neck; (c) bulk deformation by plastic flow after yield or during creep. In bonds between dissimilar materials these mechanisms will probably occur at different rates on each side of the interface.

Table 6.1 Sources, sinks and routes of mass transfer operating during diffusion bonding

Route		*Source*	*Sink*
1.	Diffusion along the surface	Surface	Bond-line neck
2.	Diffusion in the bulk	Surface	Bond-line neck
2(a).	Vapour transport	Surface	Bond-line neck
3.	Diffusion along the bond interface	Bond interface	Bond-line neck
4.	Diffusion in the bulk	Bond interface	Bond-line neck
5.	Plastic flow after yield	General asperity deformation	
6.	Creep	General asperity deformation	

void volume is accompanied by a reduction of void surface energy which is a further driving force. If we consider a cylindrical interface void it will experience a closure stress defined by the Gibbs–Thomson relation $P = g/r$ where γ is the mean void surface energy and r is its radius at a given point. The chemical potential μ' at any point on or near the interface will be changed by the local stress σ which will depend on the relative local importance of the above pressures such that $\Delta\mu' = \sigma\Omega$ where Ω is the atomic volume. Thus a gradient in the chemical potential will exist creating diffusive fluxes to close voids. However, in parallel with these chemical potential gradients, which will be identical for all components in the bonding system, are other gradients generated by geometrical and chemical differences along the joint. These will drive component diffusive fluxes in parallel with the bonding mechanisms and may interfere with or assist them.

A number of additional competing mechanisms can occur during the bonding of dissimilar materials such as metals and ceramics. These are all generated by the initial non-equilibrium nature of the interface. They may operate over a short period of time until a new equilibrium interface is formed, or if there is a chemical reaction between the two surfaces no equilibrium may be achieved and instead an extensive interfacial reaction may occur (see section 6.2.2 for more details of this). Only a few such metal–ceramic couples have been investigated, and only the Nb–Al_2O_3 system has been extensively studied [15,16]. When Nb–Al_2O_3 couples are fabricated the Al_2O_3 dissociates and aluminium and oxygen enter into solution in niobium; this is energetically favourable [17]. Burger and Rühle [15] found that in this system there was a major contribution to void closure from the dissolved aluminium and oxygen. On cooling an Nb–Al_2O_3 couple from bonding temperature the aluminium and oxygen solution in niobium became supersaturated and Al_2O_3 precipitated at favourable sites. Voids at the Nb–Al_2O_3 interface were favoured and the precipitation caused some additional void closure. In contrast, other diffusion mechanisms between dissimilar materials could be deleterious. For example, if there are significantly different diffusion rates on each side of the interface, Kirkendal porosity may be produced where material transport across the interface is faster from S_1 to S_2 than from S_2 to S_1, and to conserve matter the interface must move to accommodate the net flux. Thus the interface may move away from the voids during bonding, leaving them isolated from the fast diffusion paths. One such non-bonding mechanism – diffusion along a void surface driven by local differences in surface curvature – also occurs in metal diffusion bonds [8]. Such surface mass transfer leads to an increase in bonded area but no decrease in void volume. Also, the surface curvature at the interface is reduced, thus decreasing the driving force for other bonding mechanisms.

Clearly the modelling of metal–ceramic diffusion bonding is intrinsically a highly complex operation and a complete model encompassing all

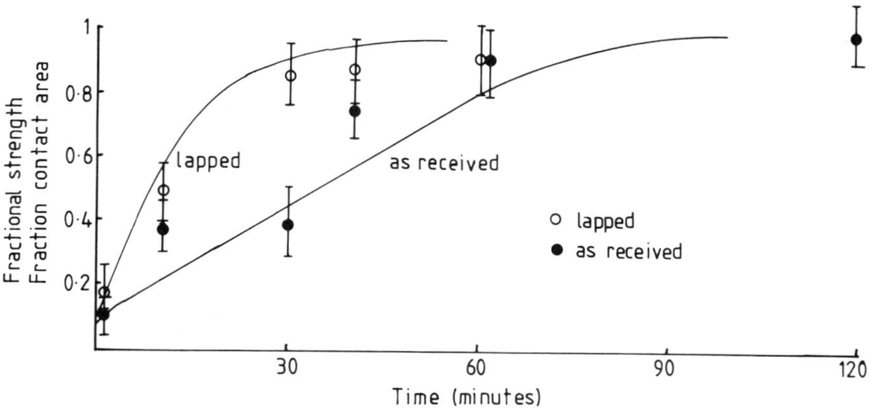

Fig. 6.4 Predictions of diffusion bonding in Al–Al_2O_3 joints. When equations derived in ref. 18 are used, an encouraging agreement between prediction and experimental results is achieved. The solid curves show the predicted area of bonding with time at 600 °C and 50 MPa pressure for the two surfaces shown in Fig. 6.1. The experimental points show the strength of the bond as a fraction of the maximum strength achieved after several hours at bonding temperature.

interactions may not be analytically possible. However, it is possible to apply simple models to certain metal–ceramic couples which show a minimum of possible complication. One such system, Al–Al_2O_3, has been investigated [18] and simple void closure models are found to be applicable in this case (Fig. 6.4). Much work needs to be done on this problem, and a major obstacle is the lack of good data on mass transfer in metal–ceramic systems.

6.2.2 Chemical Reactions

When two dissimilar materials are held in contact at an elevated temperature many chemical interactions are possible [25]. One material may take some of the other into solution, a reaction may occur to form a third phase or a reaction may occur involving the environment. In metal–ceramic bonding the situation is even more complicated because the ceramic will probably consist of at least two chemical species. Consider some of the reactions possible at an interface between a pure metal M and a simple oxide AO [17,19] (square brackets denote 'solution', i.e. [M] indicates a solution of M):

$M/AO \rightarrow M/([M] + AO)$	solution of M in AO
$M/AO \rightarrow ([A] + [O] + M)/AO$	dissociation of AO and solution of A and O in M

M/AO → ([A]/M/MO)/AO	reduction of oxide by metal and formation of a solution of A in M and MO
M/AO → M/AM/AO + O ↑	formation of an intermetallic and evolution of oxygen
M/AO + O ↓ → M/MAO_2/AO	reaction with the atmosphere to form an intermediate mixed oxide

These are just a few of the possible reactions which can occur during bonding and which have been observed in metal–ceramic couples. If the metal or ceramic contains significant alloy or grain boundary components a further effect could be the segregation of these materials towards or away from the interface.

The nature of metal–ceramic interface chemistry is covered in Chapter 7, but here we shall review the influence of bonding atmosphere on the bond strength between some metals and oxide ceramics. Gold has been very successfully bonded to oxide ceramics [20]. The nature of the chemical reaction at this interface is unclear, but an oxidizing atmosphere promotes bonding and produces a stronger joint [21]. Platinum and palladium also bond well to oxide ceramics, and a reaction has been observed in the electron microscope [22]. However, in this case the reaction is promoted by the presence of a reducing atmosphere [23]. The reaction between platinum group metals and oxide ceramics is now believed to be caused by a reduction of the oxide to form a solution of oxygen and the metal component in platinum, and a similar reaction has been proposed in the case of bonds between Al_2O_3 and niobium made in a vacuum [15,16]. The occurrence of a bulk metal–ceramic reaction can be deleterious. When nickel is bonded to Al_2O_3 in an oxidizing atmosphere an intermediate mixed oxide with a spinel structure is formed [24]; however, when these materials are bonded in a vacuum or in reducing conditions no spinel is formed and a stronger bond is produced [21]. An intermediate reaction layer is found between niobium and Al_2O_3 when they are bonded in air [26], but when they are bonded in a vacuum [16] an atomically sharp interface is observed and a stronger bond results. A similar observation has been made for ZrO_2–Ni, with the stronger bond again being made in a vacuum [10].

Most of these observations (the exception being gold Al_2O_3) are in agreement with models of metal–ceramic interfaces (e.g. Naidich [27], McDonald and Eberhart [28]) developed to explain the wetting of oxide ceramics by liquid metals. In Naidich's model a very thin interfacial zone of a mixed oxide is postulated and the work of adhesion is seen to correlate linearly with the free energy of oxidation of the metal [27]. With certain metals (e.g. magnesium, nickel and copper) a mixed spinel structure oxide may be obtained in reaction with Al_2O_3 if a source of oxygen is present, in which case the work of adhesion is now determined by the free energy of spinel formation. It is reasonable to assume that a very similar interface model is applicable to metal–ceramic solid state interfaces.

In general, if an intermediate reaction phase is formed some local volume change will occur. This volume change will generate an interfacial residual stress after bonding, and such stresses can be deleterious above some critical extent of reaction. This explains the observations from the Al_2O_3–Nb and ZrO_2–Ni systems that the bonds which had no detectable reaction layers proved to be stronger. However, in some cases a reaction has been observed in systems capable of forming strong bonds, such as noble metal–oxide bonds, but this has only been detected in experiments which were not performed under conditions which promoted true bonding and so these results are inconclusive [22]. There may be two classes of bond: some where some chemical reaction is *necessary* for adhesion, and others where bonding is intrinsic and further chemical reaction induced by the environment can result in reduced strength from residual stresses. The

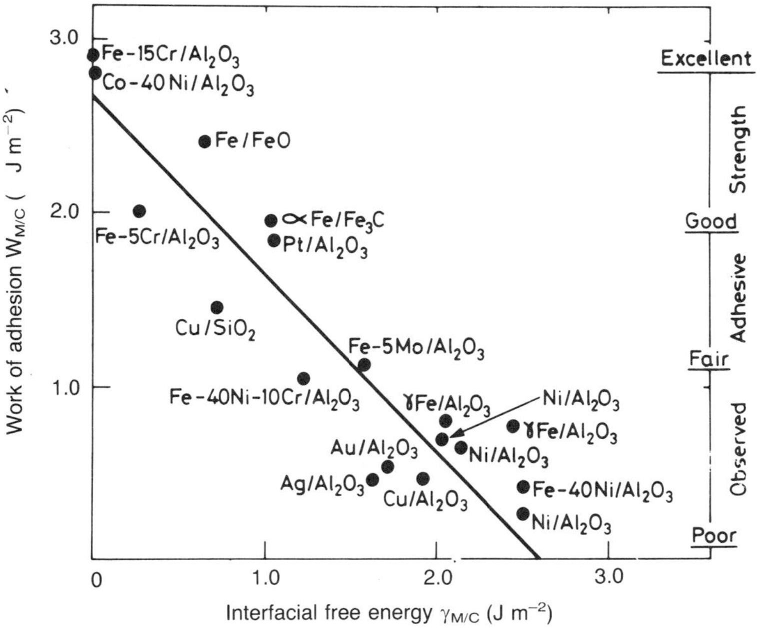

Fig. 6.5 Correlations between the thermodynamic work of adhesion, the metal–ceramic interfacial energy and the observed bond strength. There appears to be some correlation between a high work of adhesion and good (qualitative) bond strength. (After Hondros [29].)

intrinsic strength of a joint should be related to the work of adhesion, and this is qualitatively confirmed by the relation in Fig. 6.5 [29] which shows that small changes in material composition can have a substantial effect on bonding behaviour.

Interface reactions are more common in bonds to non-oxide materials, e.g. SiC or Si_3N_4. These ceramics are in general less stable with respect to metals at high temperatures. This can be exploited in the case of Si_3N_4 where bonding is promoted by the formation of eutectic metal–silicide melts with iron and nickel at comparatively low temperatures [30,31]. These reactions produce complex interfacial microstructures and are dealt with in greater detail in Chapters 5 and 10. Similar reactions have been reported in bonds between metals and SiC [32]. Strong diffusion bonds to nitrides and carbides have not been fabricated – typical bond strengths are in the region of 50–100 MPa [31]; however, this may be due to a residual stress problem, given the very low thermal expansion coefficients of these non-oxide ceramics.

6.3 THE USE OF INTERLAYERS

Ceramics and metals have quite different ranges of physical properties as illustrated by those tabulated in Chapter 1, and many problems are caused by these incompatibilities. If a joint between two elastically dissimilar materials is loaded, a local stress concentration will result [33]. In addition, diffusion bonds are fabricated at high temperatures, and on cooling internal stresses will be generated by differences in coefficients of thermal expansion. These stresses can be very large, in excess of the metal yield stress close to the interface, and can induce cracking in the ceramic or substantially reduce bond strengths. In systems where an annealing cycle is possible (e.g. Al–Al_2O_3) bond strength can be considerably increased by relieving these stresses [18].

Compliant interlayers can be used to reduce mismatched contraction stresses. The expansion strain is accommodated by a combination of elastic and plastic deformation within the interlayer, leaving a very low residual stress. Stainless steel and other metals have been successfully joined to alumina using aluminium as a ductile interlayer [9,34]. Gold has been used as an interlayer to bond a number of oxide ceramics and glasses to ferrous alloys although the joint failed if there was a very large contraction mismatch, e.g. fused SiO_2 to steel joints [14]. Another technique uses an interlayer intermediate between the ceramic and metal to reduce the interface strain. This technique is identical in concept with the graded seal long familiar in glass–metal bonding. Suganuma *et al.* [35] took this to its logical conclusion by fabricating a composite interlayer of graded composition which had contraction characteristics varying across its width from the value of the ceramic to that of the metal. Clearly, this is a complicated technique unsuited to many applications. Finite-element calculations [36] indicate that a simple interlayer with an intermediate coefficient of thermal expansion can substantially reduce residual stresses if it is sufficiently thick, even if no plastic flow occurs.

6.4 BOND STRENGTH AND TOUGHNESS

The mechanical characterization of a metal–ceramic joint is a complex problem. Even if there are no residual stresses from the bonding process, the difference in elastic properties on each side of a metal–ceramic joint will lead to interfacial stress concentrations in the absence of a defect. Once a crack is initiated and failure commences the different strain energy release rates in the metal and ceramic components can lead to deflections of the crack away from the interface. Therefore it may be necessary to adopt a component-based approach to bond strength analysis where many factors of design, and not just the mechanical properties of the bond, are considered.

The ASTM F19 standard [37] for testing brazed metallized ceramic components has been used by some workers for characterizing bond strengths [9,17,34]. However, strength is an unsatisfactory measure for brittle materials because of their notch and crack sensitivity. Hence, Elssner and co-workers [33–41] have attempted to develop fracture mechanism techniques to address this problem. This work has allowed estimates to be made of interfacial fracture energies for a number of bonded systems as described in some detail in Chapter 8. The toughness of metal–ceramic diffusion bonds is in general poor (Table 6.2) but is no worse than that of comparable brazed joints [42]. The measurement of joint toughness is difficult, and there is now ample evidence of the importance of using crack-length-independent toughness measures [42,43] which allow the prior introduction of an atomically sharp notch by wedge

Table 6.2 Metal–ceramic joint toughnesses

Joint	K_c *(MPa* $m^{1/2}$*)*	*Reference*
ZrO_2–Pd diffusion bond	7.8	10
ZrO_2–Ni diffusion bond	7.1	10
ZrO_2–CuAgTi braze	6.6	42
Si_3N_4–Hf diffusion bond	3.5	41
SiC–CuAgSnTi braze	5.6	42
Al_2O_3–Nb diffusion bond	1.9	41
	3.2	40
Al_2O_3–Al diffusion bond	0.8	41
Plasma-sprayed ZrO_2–Cr–NiCrAlY	1.0	41

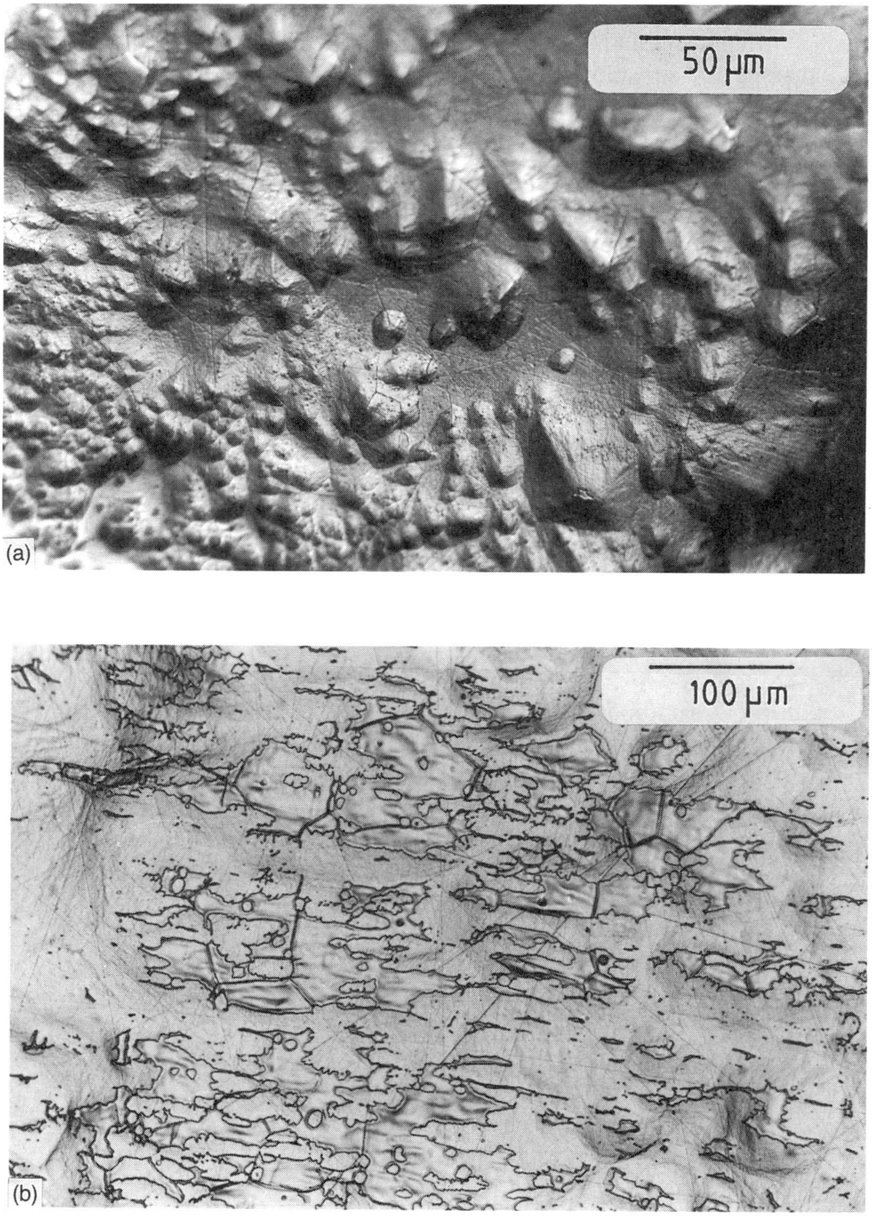

Fig. 6.6 Diffusion bonds between TZP zirconia and palladium fabricated at 1100 °C and 10 MPa pressure: (a) fracture surface of 60 min bond (palladium side), showing plastic deformation of the metal foil and the creation of numerous local bulges; (b) fracture surface of 15 min bond (palladium side) showing lower toughness than (a) with no bulging evident. The bonded regions show traces of the ZrO_2 roughness imprinted as lines.

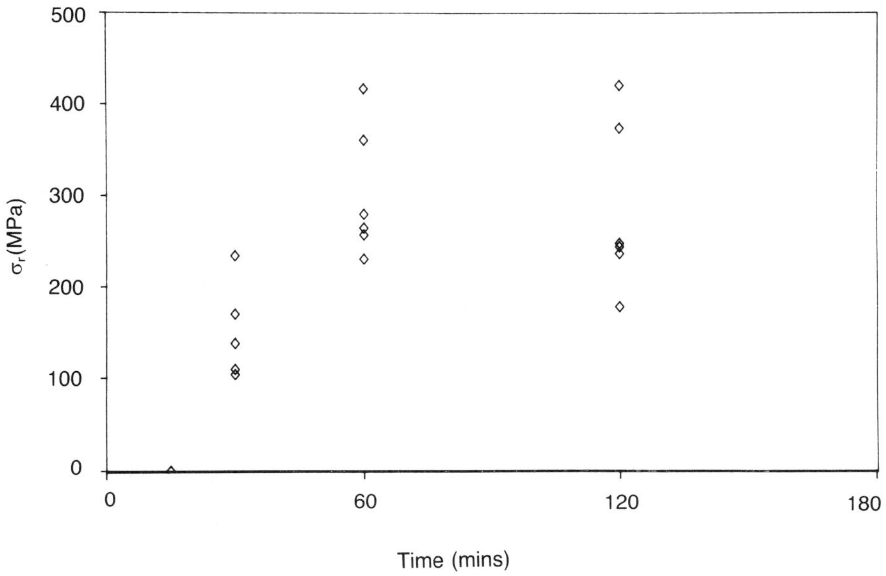

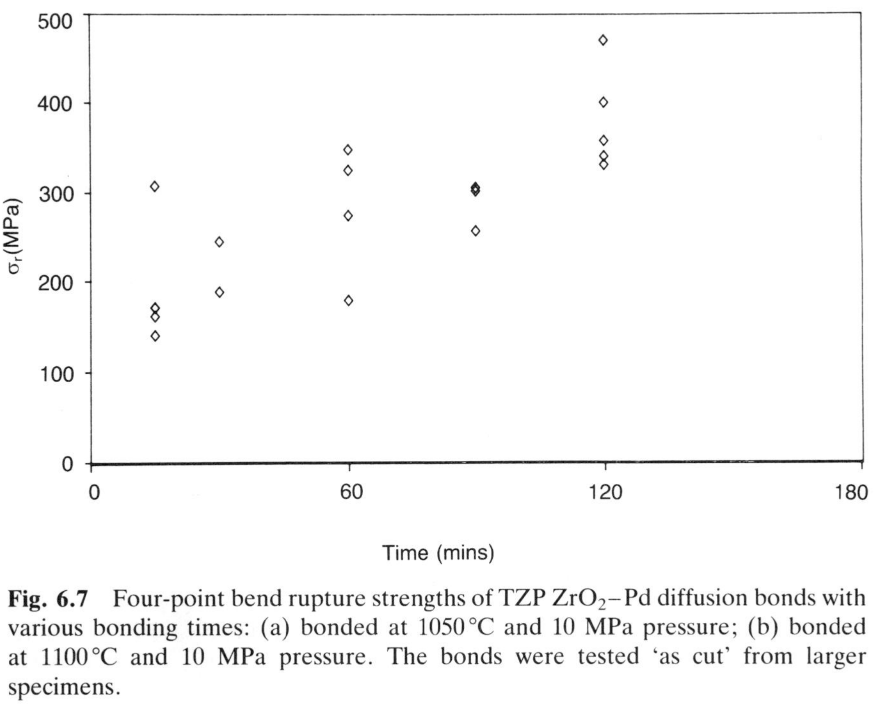

Fig. 6.7 Four-point bend rupture strengths of TZP ZrO_2–Pd diffusion bonds with various bonding times: (a) bonded at 1050 °C and 10 MPa pressure; (b) bonded at 1100 °C and 10 MPa pressure. The bonds were tested 'as cut' from larger specimens.

loading. If this is not done, fracture toughness can be significantly overestimated.

The fracture mechanisms at a metal–ceramic interface are poorly understood. In most cases fracture is reported to follow the interface, although with very good bonds fracture may start away from the joint at some critical flaw and continue through the ceramic. If a crack is initiated at the interface it will tend to follow an interfacial path and bond toughness will in the most part be governed by energy absorption in the metal. The stress field at the crack tip will exceed the yield stress of the metal, but any plastic flow will be constrained by the presence of the ceramic. Figure 6.6(a) shows the fracture surface of a Pd–ZrO_2 diffusion bond for a metal foil 0.1 mm thick sandwiched between two large ceramic blocks. The many large bulges visible are evidence for extensive plastic deformation in the metal. Figure 6.6(b) shows the palladium surface from an even weaker bond; the fracture path has closely followed one interface and an imprint of the grinding marks on the zirconia surface is visible with little evidence of metal deformation. Extensive unbonded regions are seen on both surfaces, indicating an incomplete diffusion bond.

Using Pd–ZrO_2 as an example we can see how strength and toughness evolve during bonding [44,45]. Figure 6.7 shows the variation in the mean

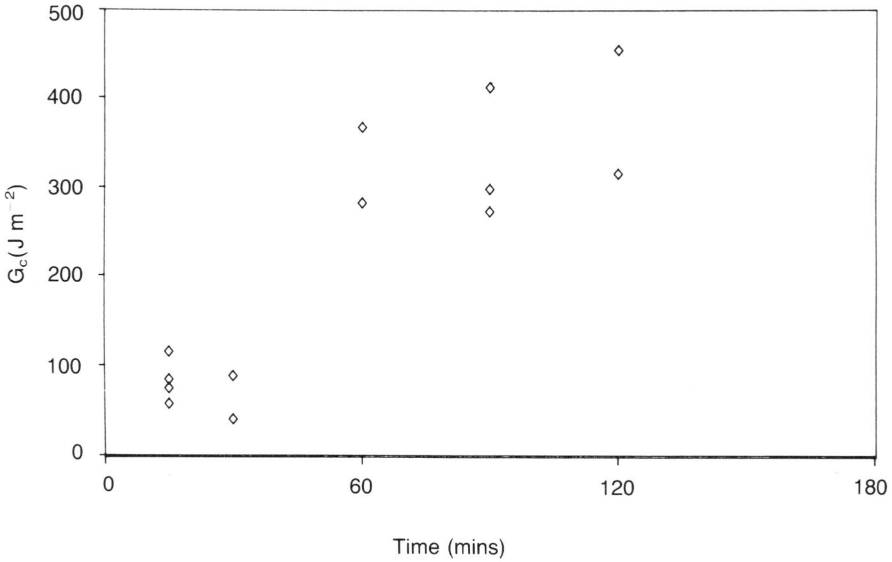

Fig. 6.8 The critical strain energy release rate G_C for fracture measured in double torsion for Pd–ZrO_2 bonds fabricated for various times at 1100 °C and 10 MPa pressure.

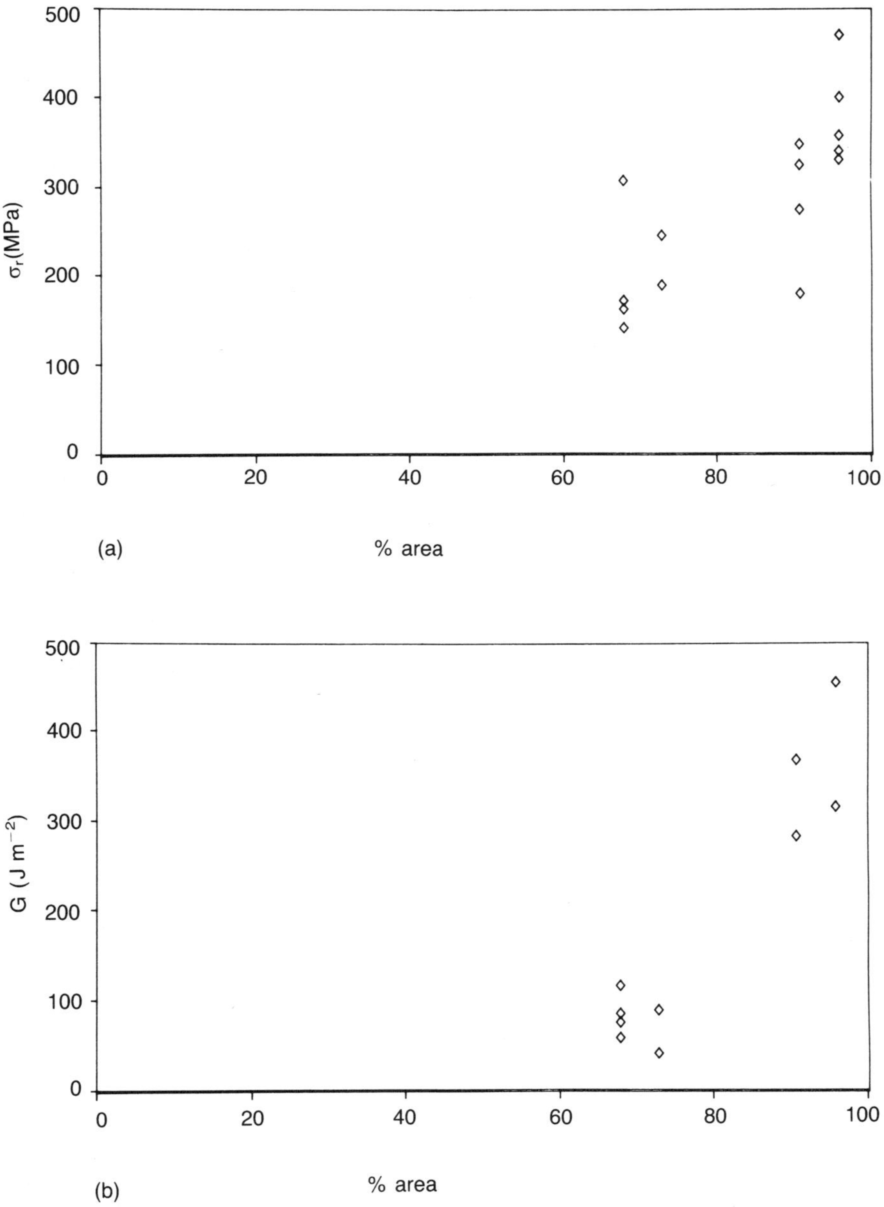

Fig. 6.9 Correlations of (a) rupture strength and (b) critical strain energy release rate with measured bonded area as determined from fracture surfaces for $Pd-ZrO_2$ bonds fabricated at 1100°C and 10 MPa pressure.

bond strength of Pd–ZrO_2 with increasing bonding time for two bonding conditions. In both cases the mean bond strength increases with time. This strength can be correlated with the degree of bonding and the area of the joint which contains interfacial voids. Figure 6.8 shows similar behaviour with bond fracture toughness measured in double torsion. The experimental scatter is such that it is difficult to determine whether either bond strength or toughness vary linearly with the bonded area (Fig. 6.9). Turwitt *et al.* [40] suggest that there is a linear relation with toughness as measured by a critical strain energy release rate G_c, but with only three points on their plot there is little conclusive evidence.

The variation in fracture surface morphology with time shown in Fig. 6.6 suggests that there is a transition in intrinsic interface behaviour. This could be interpreted in conjunction with Fig. 6.8 and 6.9 to indicate a change in fracture process with a larger G_c per bonded area at higher temperatures or longer times. This would contradict the suggestion of Turwitt *et al.* [40], but in both their work and the study of Pd–ZrO_2 diffusion bonds [44,45] the experimental scatter is too great to make an adequate conclusion as to the relation between bond toughness and degree of bonding.

6.5 CONCLUSION

Diffusion bonding is a viable technique for metal–ceramic joining which offers the same promise as reactive metal brazing. Because a solid phase reaction takes place, it is in principle suitable for higher temperature applications than are possible with braze alloys. Bond strengths are potentially very good in joints between metals and oxide ceramics, but less so with nitrides and carbides because of thermal expansion incompatibilities. Bond fracture toughness is a more serious problem, and until the mechanisms of failure are better understood little progress can be made in this area.

REFERENCES

1. Singer, G.W., Holmyard, E.J. and Hall, A.R. (1958) *A History of Technology*, Vol. 1, Oxford University Press, Oxford.
2. Becher, P.F., Halen, S.A., McDonough, W.J. and Wu, C.C. (1978) *Joining of Si_3N_4 and SiC Ceramics via Solid State Bonding*, Progress Report, Naval Research Laboratories.
3. Elssner, G., Diem, W. and Wallace, J.S. (1981) Microstructure and mechanical properties of metal to ceramic and ceramic to ceramic joints, in *Surfaces and Interfaces in Ceramic and Ceramic/Metal Systems* (eds J.A. Pask and A.G. Evans), Plenum, New York.
4. Watson, H.H.H. (1988) Culham Laboratory, personal communication.
5. Bones, R.J. (1985) Thermocompression bonding, in *Sodium–Sulfur Battery*

(eds J.L. Sadworth and A.R. Tilley), Chapman and Hall, New York.
6. Derby, B., and Wallach, E.R. (1982), Theoretical model for diffusion bonding, *Met. Sci.*, **16**, 49.
7. Derby, B. and Wallach, E.R. (1984) Diffusion bonds in copper, *J. Mater. Sci.*, **19**, 3140.
8. Derby, B. and Wallach, E.R. (1984) Diffusion bonds in iron and a low alloy steel. *J. Mater. Sci.* **19**, 3149.
9. Nicholas, M.G. and Crispin, R.M. (1987) Diffusion bonding ceramics with ductile interlayers, in *Diffusion Bonding* (ed. R. Pearce), p. 173, Cranfield.
10. Derby, B. (1987) Zirconia/metal diffusion bonds, in *Diffusion Bonding* (ed. R. Pearce), Cranfield, p. 195.
11. King, W.H. and Owczarski, W.A. (1968) Additional studies on the diffusion welding of titanium, *Weld. Res. Suppl.*, **47**, 446.
12. Garmong, G., Paton, N.E. and Argon, A.S. (1975) Attainment of full interfacial contact during diffusion bonding, *Metall Trans. A*, **6**, 1269.
13. Pilling, J., Livesey, D.W., Hawkyard, J.B. and Ridley, N. (1987) *Met. Sci.*, **18**, 117.
14. Wallach, E.R. and Hill, A. (1987) Modelling of diffusion bonding, in *Diffusion Bonding* (ed. R. Pearce), Cranfield, p.9.
15. Burger, K., and Rühle, M. (1987) Material transport mechanisms during the diffusion bonding of niobium to Al_2O_3, *Proc. Am. Ceram. Soc.*, 87th Meet., in press.
16. Burger, K., Mader, W. and Rühle, M. (1987) Structure chemistry and diffusion bonding of metal/ceramic interfaces, *Ultramicroscopy*, **22**, 1.
17. Klomp, J.T. (1987) *Stud. Phys. Theo. Chem.*, **48**, 3.
18. Derby, B. (1986) The effect of surface roughness on metal–ceramic diffusion bonding, in *Ceramic Microstructures '86: Role of Interfaces* (eds J.A. Pask and A.G. Evans), p. 319, Plenum, New York.
19. Klomp, J.T. and Vrugt, P.J. (1981) Interfaces between metals and ceramics in *Surfaces and Interfaces in Ceramic and Ceramic–Metal Systems* (eds J.A. Pask and A.G. Evans), p. 97, Plenum, New York.
20. De Bruin, H.J., Moodie, A.F. and Warble, C.E. (1978) Ceramic–metal reaction welding, *J. Mater. Sci.*, **7**, 909.
21. Bailey, F.P. and Black, K.J.T. (1978) The effect of ambient atmosphere on the gold to aluminium solid state reaction bond, *J. Mater. Sci.*, **13**, 1606.
22. Moodie, A.F. and Warble, C.E. (1977) Direct observation of Pd/MgO and Pd/SiO_2 reactions in the transmission electron microscope, *Philos. Mag.*, **35**, 201.
23. Ott, D. and Raub, C.J. (1976) The affinity of the platinum metals for refactory oxides, *Platinum Met. Rev.*, **20**, 79.
24. Calow, C.A. and Porter, I.T. (1971) The solid state bonding of nickel to alumina, *J. Mater. Sci.*, **6**, 156.
25. Klomp, J.T. (1980) Interfacial reaction between metals and oxides during sealing, *Ceram. Bull.*, **59**, 794.
26. Morozumi, S., Kikuchi, M. and Nishino, T. (1981) Bonding mechanisms between alumina and niobium, *J. Mater. Sci.*, **16**, 2137.
27. Naidich, J.W. (1981) *Prog. Surf. Membr. Sci.*, **14**, 353.
28. McDonald, J.E. and Eberhart, J.G. (1965) Second Inter. cont. on science of hard materials, *Trans. Metall. Soc. AIME*, **233**, 512.
29. Hondros, E.D. (1984) Bonding of metal/ceramic interfaces, Adhesion in aluminium oxide – metal systems, *Inst. Phys. Conf. Ser.*, **75**(2), 121.
30. Suganuma, K., Okamoto, T., Miyamoto, Y., Schimada, M. and Koizumi, M.

(1986) Joining Si_3N_4 to type 405 steel with soft metal interlayers, *Mater. Sci. Technol.*, **2**, 1156.
31. Suganuma, K., Okamoto, T., Koizumi, M., Fujiata, T. and Niihara, K. (1987) Joining of silicon nitride with metallic interlayers. In *Conf. on High Technology Joining* (ed. M.G. Nicholas) BABS, London, p. 101.
32. Morozumi, S., Endo, M., Kikuchi, M. and Hamajima, K. (1985) Bonding mechanism between silicon carbide and thin foils of reactive metals, *J. Mater. Sci.*, **20**, 3976.
33. Evans, A.G. and Rühle, M. (1985) On the mechanics of failure in ceramic/metal bonded systems, *Mater. Res. Soc. Symp. Proc.*, **40**, 153.
34. Nicholas, M.G. and Crispin, R.M. (1982) Diffusion bonding stainless steel to alumina using aluminium interlayers, *J. Mater. Sci.*, **17**, 3347.
35. Suganuma, K., Okamoto, T., Shimada, M. and Koizumi, M. (1983) New method for solid-state bonding between ceramics and metals, *Commun. Am. Ceram. Soc.*, **66**, C117.
36. Suganuma, K., Okamoto, T., Koizumi, M. and Shimada, M. (1984) Effect of interlayers in ceramic–metal joints with thermal expansion mismatches, *Commun. Am. Ceram. Soc.*, **67**, C256.
37. *Standard Method for Tension and Vacuum Testing Metallized Ceramic Seals*, ASTM F19-64, 1964.
38. Elssner, G., Suga, T. and Turwitt, M. (1985) Fracture of ceramic to metal interfaces, *J. Physi. (Paris) C*, **4**, 597.
39. Suga, T. and Elssner, G. (1985) Characterisation of the adherence of metals by a fracture mechanics concept, *MRS Symp. Proc.*, **40**, 203.
40. Turwitt, M., Elssner, G. and Petzow, G. (1986) On the fracture behaviour and microstructure of metal-to-ceramic joints. In *Ceramic Microstructures '86: Role of Interfaces* (eds J.A. Pask and A.G. Evans) Plenum, New York, p. 969.
41. Suga, T. and Elssner, G. (1985) Determination of the fracture energy and the fracture resistance of interfaces', *J. Physi. (Paris) C*, **4**, 657.
42. Moorhead, A.J. and Becher, P.F. (1987) Adaption of the DCB test for determining fracture toughness of brazed joints in ceramic materials, *J. Mater Sci.*, **22**, 1297.
43. Parry, T.V., Igbinechion, S.J. and Wronski, A.S. (1986) Use of the double torsion technique to evaluate fracture energy of adhesive cements in the temperature range 20–700 °C, *Proc. Br. Ceram. Soc.*, **37**, 281.
44. James, N.L. (1988) *Zirconia Palladium Diffusion Bonds*. Ph.D. Thesis, Department of Metallurgy and Science of Materials, University of Oxford.
45. James, N.L. and Derloy, B. (1989) Zirconia Palladium Diffusion Bonds, in *Joining Ceramics, Glass and Metal* (ed. W. Kraft) DGM, Oberusel FRG, p. 73.

7

Physicochemical aspects of ceramic–metal joining

J.T. Klomp

7.1 INTRODUCTION

In this chapter we review the physicochemical background of the ceramic–metal interactions needed for the formation of an adherent interface. The application of equilibrium thermodynamics is considered since this provides information on the conditions that have to be achieved in a ceramic–metal system for an interface reaction to proceed. Also, some physical aspects of the materials that influence the structural stability of a ceramic–metal joint are considered. The first step in creating an adherent interface is to achieve intimate contact, and thereafter bonding can occur. However, if the bonding is to be useful, it must be able to endure the stresses and strains to which interfaces are subjected after fabrication. The first and second of these steps – contact and bonding – will be considered in some detail, and a few concluding remarks about joint stability will then be made.

7.2 INTERFACE FORMATION

The driving force for the formation of an interface between materials is the energy decrease of the system resulting from its establishment. The interfacial energy should reach the lowest achievable value as the bond is formed, otherwise further changes that could degrade the stability of the bond may occur under operating conditions.

Interaction between the ceramic and the metal requires the surfaces to be in direct intimate contact, and hence the surfaces must be free of contaminants. Thus the atmosphere is important in achieving the maximum energy decrease. In practice, formation processes are usually carried out at elevated temperatures so that the rates of interaction are rapid. In this section we consider the effect on interface formation of both physical and chemical interactions.

7.2.1 Liquid–solid systems

If contact between the ceramic and the metal is established, an interface can be formed. The driving force for the interface formation is the energy W_{ad} which the system gives up, and this is expressed by the Dupré equation:

$$W_{ad} = \gamma_{SV} + \gamma_{LV} - \gamma_{SL}$$

where W_{ad} is the reversible work of adhesion, i.e. the work required to separate a unit area of the interface into the two original surfaces, and the γ terms are the energies of the solid surface, the liquid surface and the solid–liquid interface. For a solid ceramic–liquid metal system this can be rewritten as

$$W_{ad} = \gamma_c + \gamma_m - \gamma_{cm} \qquad (7.1)$$

Direct measurement of W_{ad} or γ_{cm} is not possible, and so in practice W_{ad} is derived by measuring the equilibrium contact angle u, illustrated in Chapter 2, Fig. 2.5, which a solid or liquid metal establishes in contact with a ceramic.

The Young equation for the system is

$$\gamma_c - \gamma_{cm} - \gamma_m \cos\theta = 0 \qquad (7.2)$$

and by combining equations (7.1) and (7.2) we obtain

$$W_{ad} = \gamma_m (1 + \cos\theta) \qquad (7.3)$$

Values for γ_m are readily available in the literature, and hence W_{ad} can be derived.

It should be noted that the derivation of W_{ad} from contact angle measurements is only valid if γ_m and γ_c do not change during the experiment, which means that chemical reactions should not have occurred. A description of W_{ad} in physical terms is outside the scope of this chapter but details can be found in the literature [1]. A more extensive treatment of wetting aspects is given in Chapter 2.

7.2.2 Solid–solid systems

If neither component of a ceramic–metal system is in the liquid state during the joining process, the formation of a continuous interface is less easily achieved. This is due to the much lower mobility of the atomic species in a solid. Solid state bonding involves heating the system under an applied mechanical load to establish contact between the materials. In this process the formation of a continuous interface is mainly determined by the surface quality of the ceramic. Surface irregularities introduced by machining or due to grain boundaries hamper the formation of a continuous interface. Increasing the load will increase the contact area. However, the plastically flowing metal cannot fill sharp grooves completely because of opposing metal surface energy effects. Even if the plastic deformation is extremely large, as required in bonding metals such as aluminium which are covered with stable surface oxide films [2], pores will be left at the interface. Interface pores can only be eliminated by transport of matter, e.g. metal evaporation and condensation on the ceramic, so that the interface pore becomes a pore in the metal. Such pores can only be eliminated by vacancy diffusion; the driving force is the change in the free surface area A yielding the energy $g_m dA$ and the stress gradient due to the applied load. The rate of decrease of the pore volume V_p is given by the relation

$$-\frac{dV_p}{dt} = \frac{8p\, D_v\, g_m\, V}{kT} \tag{7.4}$$

where t is the time, D_v is the vacancy diffusion coefficient, Ω is the vacancy volume, T is the temperature and k is Boltzmann's constant. As long as the diffusion coefficient of the ceramic is much lower than that of the metal, the ceramic does not make a significant contribution to the process. The diffusing vacancies can condense on the free metal surface provided that the diffusion path length l equals the distance L of the vacancy source from the free metal surface: $l = (D_v t)^{1/2} = L$. If L is greater than $(D_v t)^{1/2}$, vacancy condensation that can give rise to the formation of microvoids takes place on grain boundaries or lattice defects.

Residual surface flaws have a deleterious effect on the mechanical properties and hermeticity of the interface. Therefore smooth sintered surfaces or properly machined surfaces are preferable. If the ceramic surface is smooth, the time required for the formation of a continuous interface is determined by the deformation rate of the metal and can be much less than 1 s [2]. This is of particular advantage in bonding oxygen-active metals such as aluminium.

7.3 INTERFACIAL BONDING

Chemical reactions at the metal–ceramic interface are the key factor in adherence since they decrease further the free energy of the ceramic–metal system once intimate contact is established and physical interaction has occurred [3].

Chemical descriptions of bonding processes should be in terms of irreversible thermodynamics, but the lack of data makes this approach impractical. The kinetics of the reaction have to be determined experimentally, but guidance as to the nature of the reactions can be derived from equilibrium thermodynamic data. The driving force for the reaction is the chemical potential of the species involved, which causes charge transport across the interface. Usually, but not always, mass transport also occurs, which can lead to the formation of interfacial layers.

7.3.1 Metal–non-metal reactions

Thermodynamic data on reactions between metals and non-metals such as oxygen are available in the literature [4–6]. Using Ellingham diagrams, such as Fig. 7.1, a first indication can be obtained as to whether a metal in contact with a ceramic will react with its non-metallic component. Similar diagrams for complex oxides, e.g. silicates, and for nitrides and carbides

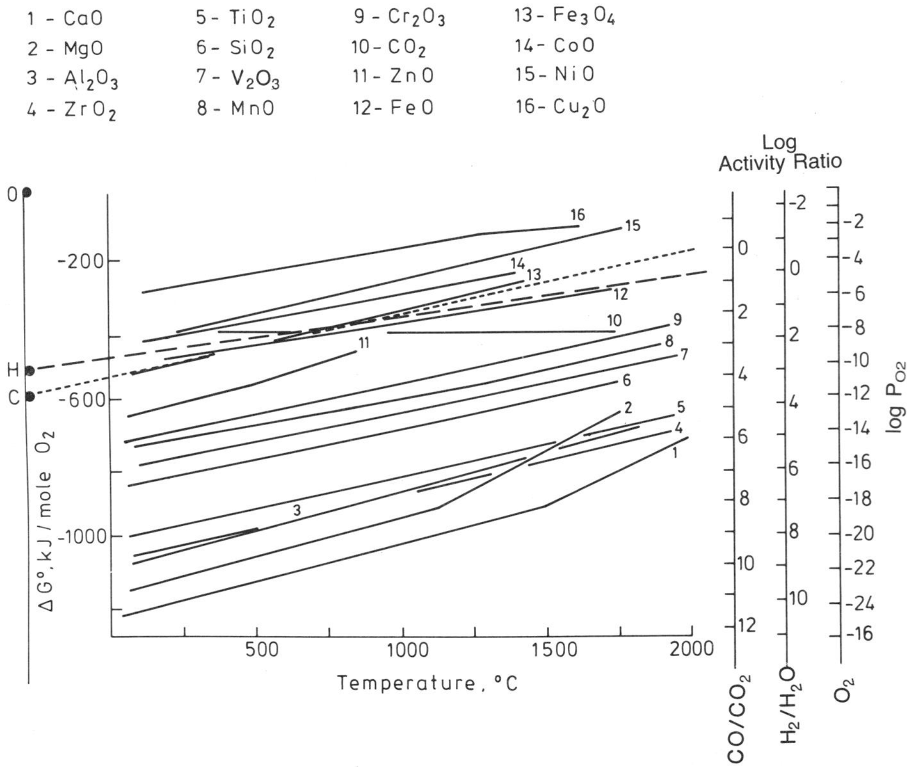

Fig. 7.1 The standard free energies of formation of oxides per gram-mole of oxygen (gas). (5)

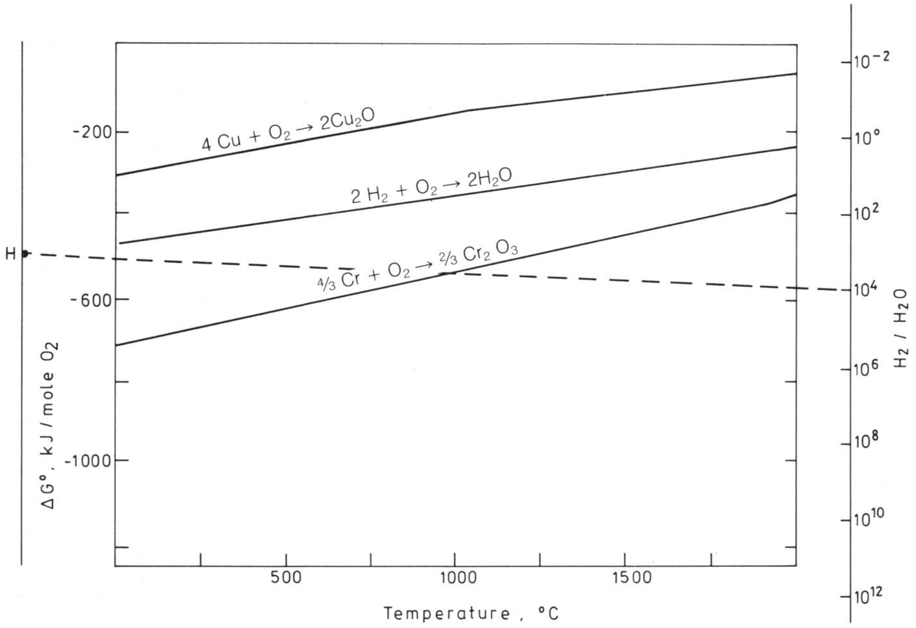

Fig. 7.2 Part of fig. 1

are available in the literature [4,5] or can be constructed from available thermodynamic data [6].

Figure 7.1 gives the affinity of a number of elements for oxygen as a function of temperature. For clarity Fig. 7.2 includes only the lines for $2Cu_2O$, $2H_2O$ and $\frac{2}{3}Cr_2O_3$. These represent the Gibbs free energy ΔG° per gram mole O_2 of the equilibrium oxidation reactions of copper, hydrogen and chromium at standard pressure (1 bar O_2):

$$4\,Cu + O_2 \rightarrow 2Cu_2O \quad \Delta G^\circ_{F(Cu_2O)} = f(T) \tag{7.5}$$

$$2H_2 + O_2 \rightarrow 2H_2O \quad \Delta G^\circ_{F(H_2O)} = f(T) \tag{7.6}$$

$$\tfrac{4}{3}Cr + O_2 \rightarrow \tfrac{2}{3}Cr_2O_3 \quad \Delta G^\circ_{F(Cr_2O_3)} = f(T) \tag{7.7}$$

The value of ΔG° can also be derived using tables of thermodynamic data [6] and the formula

$$\Delta G^\circ_F = \Delta H^\circ - T\Delta S^\circ \tag{7.8}$$

The table referred to give G_F, H and S for the reactants and the reaction products, from which the ΔG°, ΔH° and ΔS° terms can be calculated.

Comparing the position of the lines in Fig. 7.2 we see that ΔG_F° for Cr_2O_3 is more negative than for H_2O, which in turn is more negative than ΔG_F° for Cu_2O, showing that the chemical stability of an oxide increases with decreasing ΔG_F°. Thus, at standard temperature and oxygen pressure, Cu_2O will be reduced by hydrogen, giving copper and H_2O as reaction products, but Cr_2O_3 will not be reduced. However, Cr_2O_3 can be reduced if the conditions are changed in such a way that the ratio H_2/H_2O is much larger than that for the equilibrium reaction.

The H_2 line drawn in Fig. 7.2 is for the equilibrium reaction at 1 atm. If water vapour is removed so that the H_2/H_2O ratio is increased, the activity ratio in the following equation becomes smaller than unity:

$$\Delta G_R = \Delta G_R^\circ + RT \ln \left\{ \frac{a(H_2O)}{a(H_2)} \right\} \tag{7.9}$$

In Fig. 7.2 this can be represented by a family of lines radiating from point H at $T = 0$ K corresponding to H_2/H_2O activity and pressure ratios indicated by their intersection with the second of the right-hand ordinates. If a radiating line (dotted in Fig. 7.2) is drawn to intersect the Cr_2O_3 line at 1000 °C, this has a slope corresponding to an H_2/H_2O ratio of 10^4. Above 1000 °C ΔG_R for the H_2 line is more negative than that for Cr_2O_3 and so the oxide will be reduced. The condition for this is that the H_2O pressure must be less than 10^{-4} of the hydrogen pressure. At lower temperatures, reduction of Cr_2O_3 requires even lower H_2O pressures.

An identical argument holds if CO gas is used as a reducing agent (see Fig. 7.1). The third ordinate on the right in Fig. 7.1, pO_2, relates to the equilibrium oxygen pressure, or activity, for the reaction $n\text{Me} + O_2 \rightarrow \text{Me}_nO_2$. The standard oxygen pressure of 1 bar at 0 K corresponds to the point O on the line on the left of the diagram. This point represents the free enthalpy ΔH° of oxygen which is by definition zero. If a straight line is constructed starting from point O to interest the curve for Cu_2O at 1000 °C it will continue on to intersect the pO_2 ordinate at a value of about 10^{-7}. Hence the equilibrium oxygen pressure of the reaction $4Cu + O_2 \rightarrow 2Cu_2O$ is 10^{-7} bar. Similarly Al_2O_3 cannot be reduced by 1 bar H_2 at 1000 °C unless the partial pressure of H_2O in the hydrogen is below 10^{-10} bar. This partial pressure cannot be achieved in practice and, in fact, a value of even 10^{-6} bar is hard to obtain. According to the diagram, the reduction of Al_2O_3 by 1 bar H_2 containing 10^{-6} bar H_2O occurs only at 1800 °C and above.

It also follows from the relative positions of the lines in the diagram that an oxide can be reduced by a metal only if ΔG_F° of the oxide of the metal is more negative than ΔG_F° of the oxide, e.g. chromium reduces Cu_2O to copper by forming Cr_2O_3. This effect and the ability of metals to dissolve the metallic and the non-metallic components of a ceramic form the basis of metal–ceramic joining by chemical reactions. Thus the study of phase diagrams of metal–ceramic systems is essential to understanding joining processes. The use of equilibrium thermodynamic data can be illustrated further by reference to two technologically important systems.

(a) $Pt{-}Al_2O_3{-}H_2$

Platinum has a very low affinity for oxygen, but reactions can occur in hydrogen environments. However, hydrogen does not have to reduce the ceramic because platinum destabilizes the Al_2O_3 as a result of its great affinity for aluminium. Thus the heat of solution of aluminium in platinum is very negative, and this enables platinum to interact with Al_2O_3 in a hydrogen environment despite the thermodynamic stability of Al_2O_3 [7]:

$$\tfrac{1}{3}Al_2O_3 \rightarrow \tfrac{2}{3}Al + \tfrac{1}{2}O_2 \qquad \Delta G^\circ_R = 564190 - 108.74T \tag{7.10}$$

$$H_2 + \tfrac{1}{2}O_2 \rightarrow H_2O \qquad \Delta G^\circ_R = -250450 + 57.29T \tag{7.11}$$

$$Pt + \tfrac{2}{3}Al \rightarrow \tfrac{2}{3}[Al]_{Pt} \qquad \Delta G^\circ_R = \Delta H^\circ = -109870 \tag{7.12}$$

$$\tfrac{1}{3}Al_2O_3 + H_2 + Pt \rightarrow \tfrac{2}{3}[Al]_{Pt} + H_2O \qquad \Delta G^\circ_R = 203870 - 51.45T \tag{7.13}$$

The heat of solution ΔH° has been equated to the free energy ΔG° of reaction (7.12) from tabulated data [7] and is a good approximation for metal–metal solute solutions. However, this approximation does not hold for gases dissolved in metals; for that case a more extensive procedure is necessary as will be shown later. Since we are dealing with non-standard conditions, the free energy of the reaction changes with the activity a of the dissolved aluminium and with the H_2O pressure:

$$\Delta G_R = \Delta G^\circ_R + RT\left[\tfrac{2}{3}\ln a(Al) + \ln\left\{\frac{a(H_2O)}{a(H_2)}\right\}\right] \tag{7.14}$$

Data for activities are scarce, and so it is often necessary to introduce a simplification by assuming that the aluminium dissolves ideally in platinum, i.e. it obeys Raoult's law which implies that the activity a equals the atom fraction x_F, so that $a(Al) = x_F(Al)$. Equation (7.14) then becomes

$$\Delta G_R = \Delta G_R^\circ + \tfrac{2}{3}RT \ln \left\{ x_F(Al) + \frac{a(H_2O)}{a(H_2)} \right\} \tag{7.15}$$

The ratio $a(H_2)/a(H_2O)$ or $p(H_2)/p(H_2O)$ can now be calculated for $\Delta G_R = 0$ and various values of $x_F(Al)$, and the results are shown in Fig. 7.3. The combined influence of the entropy and of the heat of solution of aluminium in platinum on the required ratio $a(H_2)/a(H_2O)$ is illustrated by the position of these lines and the line for the reduction of Al_2O_3 by hydrogen in the absence of platinum.

These calculations indicate that the atmosphere does not have to affect the free surfaces of the ceramic but only the reaction at the ceramic–metal interface. If multicomponent ceramics have to be bonded to metal it is necessary to include ceramic decomposition in the range of reactions considered and to determine which component is the most likely to react with the metal [8]. The relevance of these calculations is indicated by experiments which have shown that pure platinum in contact with Al_2O_3 at 1800 K in 1 bar of hydrogen containing 2 ppm H_2O attacks the sapphire and the platinum is enriched in aluminium by up to 7% [9]. Similar results are observed on $Pt-ZrO_2$ interfaces formed in hydrogen [10].

These thermodynamic considerations refer to equilibrium reactions only. The actual kinetics of the reaction must be determined experimentally. Gas-assisted ceramic–metal reactions are not restricted to platinum

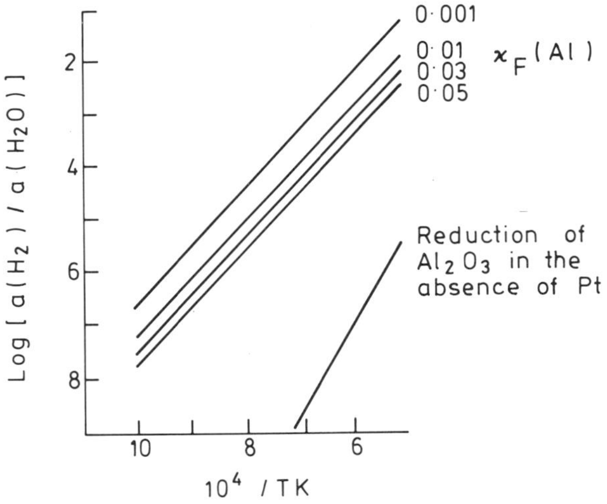

Fig. 7.3 Reaction conditions for the Pt–Al_2O_3–H_2 system. Reaction occurs below the line of a given concentration of Al in the Pt ([Al] Pt).

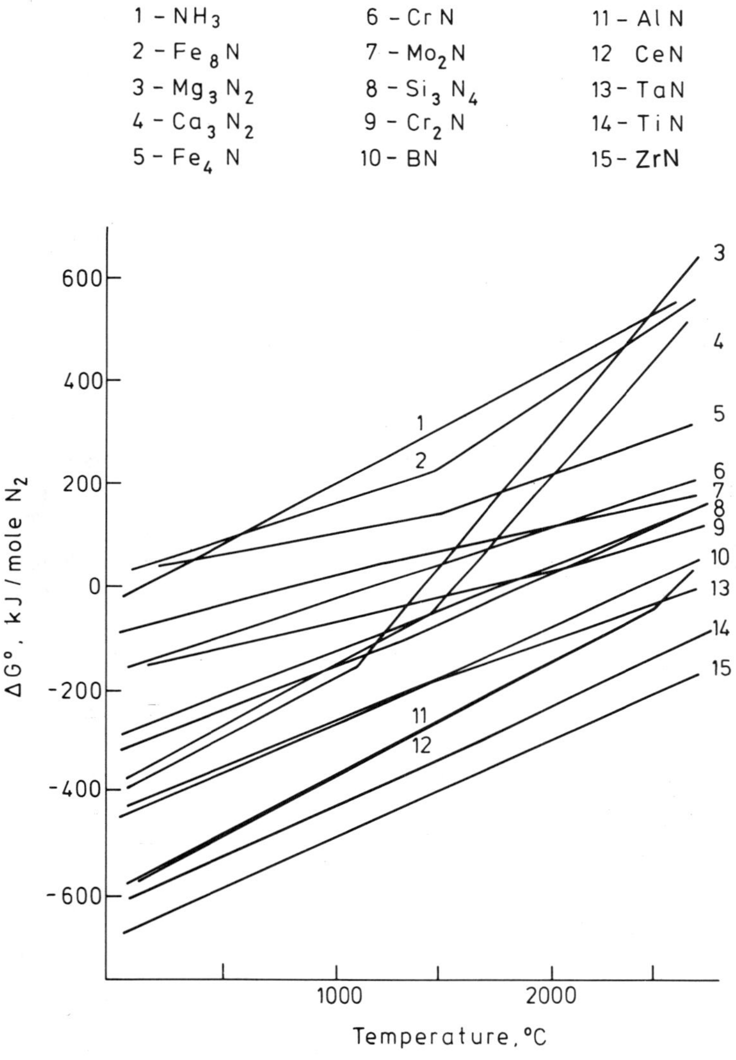

Fig. 7.4 The standard free energy of formation of nitrides per gram–mole nitrogen (gas) (4).

but could occur with any metal having a low affinity for the non-metal element of the ceramic but a high affinity and ability to dissolve the metal species.

(b) Ti–Si_3N_4

As the second illustration of a ceramic–metal reaction, consider the Si_3N_4–Ti system in a low-pressure environment. The Ellingham diagram for some nitrides is shown in Fig. 7.4. The free energy of formation of Si_3N_4 and of TiN per mole N_2 can be represented by the equations

$$\Delta G^{\circ}_{F(Si_3N_4)} = -396480 + 206.64T \tag{7.16}$$

$$\Delta G^{\circ}_{F(TiN)} = -679140 + 191.52T \tag{7.17}$$

Comparison with the free energy of oxide formation for silicon and titanium (Fig. 7.1) represented by the equations

$$\Delta G^{\circ}_{F(SiO_2)} = -873600 + 178.08T \tag{7.18}$$

$$\Delta G^{\circ}_{F(TiO_2)} = -911400 + 176.82T \tag{7.19}$$

shows that the nitrides are less stable. This implies that the oxygen partial pressure in the reaction environment should be extremely low to prevent the Si_3N_4 and titanium from oxidizing. Equations (7.16) and (7.17) show that Si_3N_4 is less stable than TiN, and so TiN will form when titanium is in contact with Si_3N_4 and the silicon liberated will react to form titanium

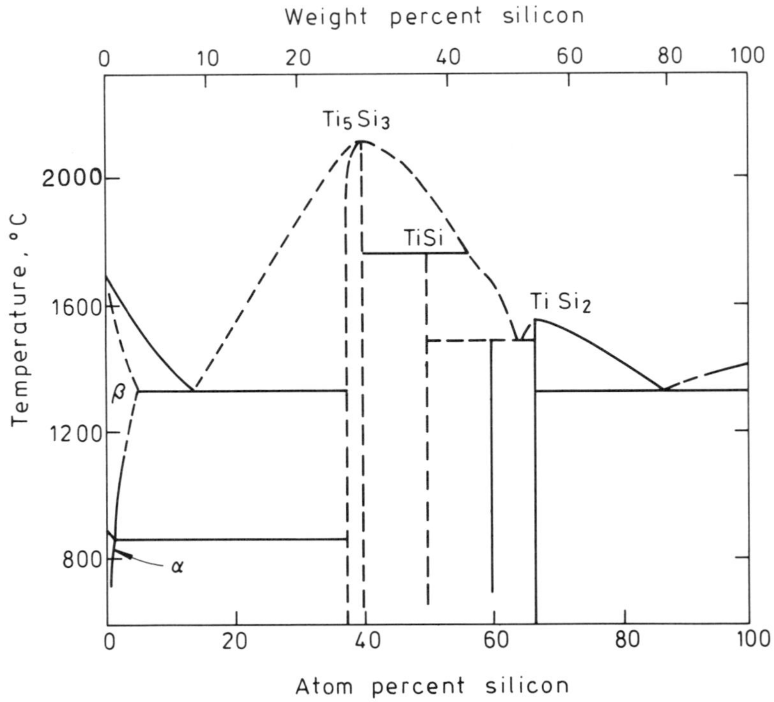

Fig. 7.5 Si–Ti phase diagram (11).

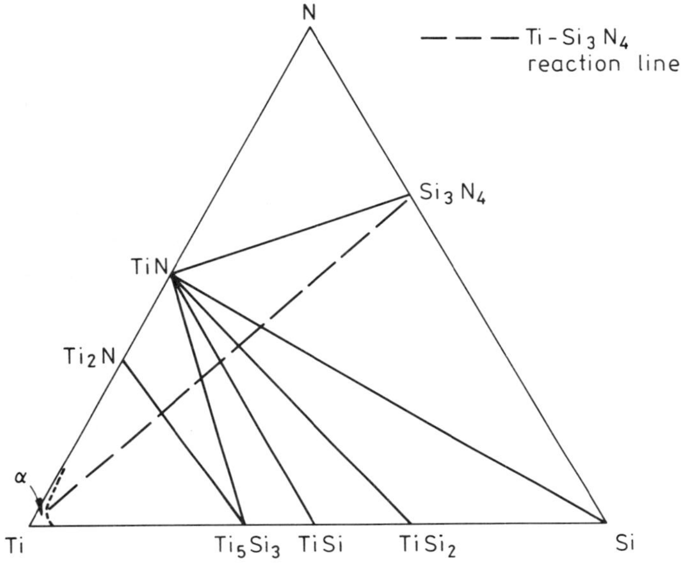

Fig. 7.6 Ti–Si–N phase diagram (12).

silicides (Fig. 7.5) [11]. Similarly the Ti–Si–N phase diagram (Fig. 7.6) shows that Si_3N_4 is only stable in the presence of TiN [12].

The conditions for the Ti–Si_3N_4 reaction in the temperature range 1000–1500 K can be calculated using the reaction equations

$$\tfrac{1}{4}Si_3N_4 \rightarrow \tfrac{3}{4}Si + \tfrac{1}{2}N_2 \qquad \Delta G^\circ_R = 198240 - 103.32T \tag{7.20}$$

$$Ti + \tfrac{1}{2}N_2 \rightarrow TiN \qquad \Delta G^\circ_R = -339570 + 95.76T \tag{7.21}$$

$$TiN \rightarrow [N]_{Ti+Ti} \qquad \Delta G^\circ_R = 55999 + 4.83T \tag{7.22}$$

$$\tfrac{3}{4}Si + \tfrac{5}{4}Ti \rightarrow \tfrac{1}{4}Ti_5Si_3 \qquad \Delta G^\circ_R = -147739 + 1.34T \tag{7.23}$$

$$\tfrac{1}{4}Ti_5Si_3 \rightarrow \tfrac{3}{4}[Si]_{Ti} + \tfrac{5}{4}Ti \qquad \Delta G^\circ_R = 37027 + 4.28T \tag{7.24}$$

$$\tfrac{1}{4}Si_3N_4 + \tfrac{9}{4}Ti \rightarrow \tfrac{3}{4}[Si]_{Ti} + [N]_{Ti} \qquad \Delta G^\circ_R = -196043 + 2.89T \tag{7.25}$$

ΔG_R° values for the solution reactions (7.22) and (7.24) are obtained using van't Hoff's equation

$$\Delta H^\circ = -R\frac{d\ln x_F}{d(\Delta G_R/T)}$$

where x_F is the concentration of nitrogen in titanium. The thermodynamic data for TiN according to equation (7.21) and the solid solution values of nitrogen in titanium with temperature (Fig. 7.7) are used to calculate ΔH°. Then ΔS° can also be obtained.

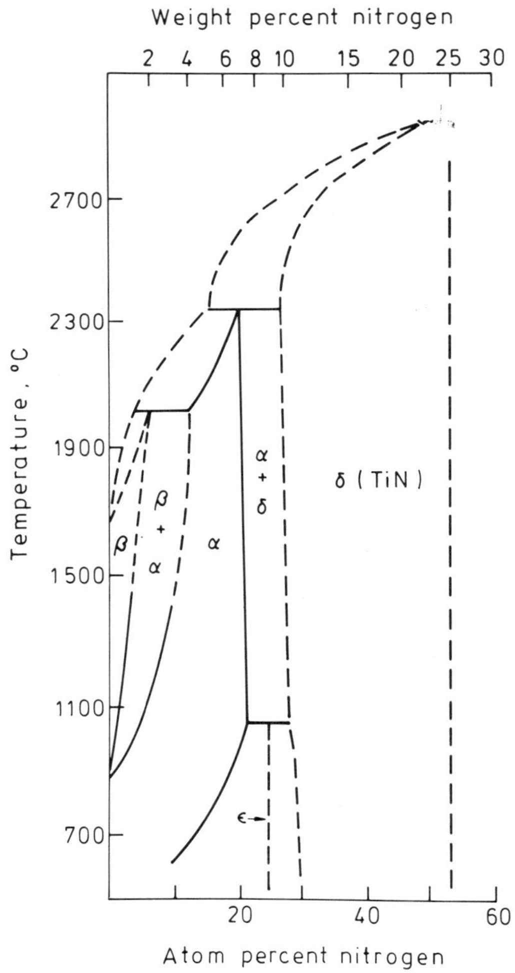

Fig. 7.7 Ni–Ti phase diagram (11).

For standard conditions we have

$$\begin{array}{ll} \mathrm{Ti} + \tfrac{1}{2}\mathrm{N_2} \rightarrow \mathrm{TiN} & \Delta G^\circ \text{ is known} \\ \mathrm{TiN} \rightarrow [\mathrm{N}]_{\mathrm{Ti}} + \mathrm{Ti} & \Delta G^\circ \text{ is unknown} \\ \hline \tfrac{1}{2}\mathrm{N_2} \rightarrow [\mathrm{N}]_{\mathrm{Ti}} & \Delta G^\circ_{\mathrm{R}} \text{ is unknown} \end{array}$$

For non-standard conditions we have

$$\Delta G_{\mathrm{R}} = \Delta G^\circ_{\mathrm{R}} + RT \ln x_{\mathrm{F}} \tag{7.26}$$

At equilibrium $\Delta G_{\mathrm{R}} = 0$, and so $\Delta G^\circ_{\mathrm{R}} = RT \ln x_{\mathrm{F}}$.

We can obtain ΔS° from $\Delta S^\circ = \mathrm{d}(\Delta G^\circ)/\mathrm{d}T$ and ΔG° from the calculated value of ΔH°. The same procedure leads to the $\Delta G^\circ_{\mathrm{R}}$ of the dissolution of silicon in titanium. Had we taken for $\Delta G^\circ_{\mathrm{R}}$ the value of ΔH° from ref. 5, which gives $[\mathrm{Si}]_{\mathrm{Ti}} = -142800\mathrm{J}$, whereas the sum of equations (7.23) and (7.24) yields $\Delta G^\circ_{\mathrm{R}} = -110712 + 5.63T$, we would see a difference of less than 30% at $T = 0$, which is quite acceptable for calculations of this type. The $\Delta G^\circ_{\mathrm{R}}$ of the total reaction (equation (7.25)) is very negative for the temperature range 1000–1500 K, and so equilibrium is not possible, in accord with the phase diagram (Fig. 7.6). The nitrogen pressure in the environment is determined by the equilibrium of N_2 with the $[\mathrm{N}]_{\mathrm{Ti}}$. This means that the behaviour of the system has to be considered for non-standard conditions. Therefore equations (7.21) and (7.22) have to be added, which yields

$$\tfrac{1}{2}\mathrm{N_2} \rightarrow [\mathrm{N}]_{\mathrm{Ti}} \qquad \Delta G^\circ_{\mathrm{R}} = -283570 + 100.6T \tag{7.27}$$

For non-standard conditions

$$\Delta G_{\mathrm{R}} = \Delta G^\circ_{\mathrm{R}} + RT[\ln x_{\mathrm{F}} - \tfrac{1}{2}\ln\{a(\mathrm{N_2})^{1/2}\}] \tag{7.28}$$

Thus the allowable partial pressures (activities) of nitrogen in the environment can be calculated for known values of x_{F} in equilibrium conditions, as shown in Fig. 7.8.

As already mentioned, equilibrium thermodynamics are only useful for determining the conditions in which a ceramic–metal system can react, but they do not say anything about the reaction kinetics. These have to be determined experimentally by using available data on the diffusion of the ceramic species into the metal and by instrumental analytical techniques. The formation of intermetallic or non-metallic compounds should be prevented as far as possible because these compounds almost always have a detrimental effect on the mechanical properties of the ceramic–metal

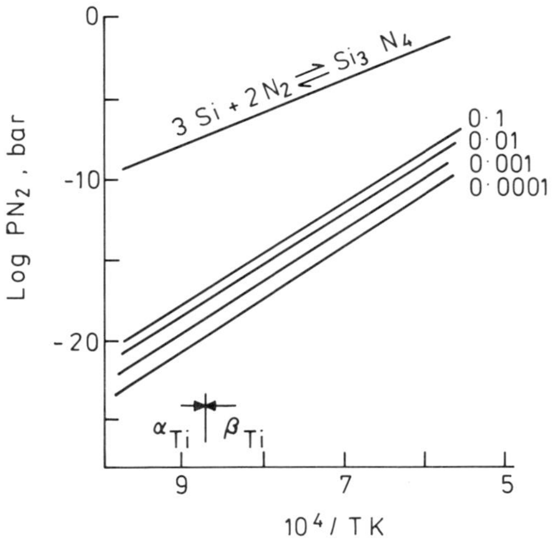

Fig. 7.8 Variation of $LogPN_2$ with (1/T) for the Si_3N_4–Ti reaction.

joint. Preferably the formation of solid solutions should be aimed at, and this requires close control of the bonding conditions.

7.4 JOINT STABILITY

It is well known that differences in thermal expansion between metals and ceramics introduce thermal stresses in a metal–ceramic joint. For designs with non-ductile materials we refer the reader to textbooks such as ref. 13. The range of thermal expansion coefficients is rather broad for oxide ceramics, but very narrow and low for non-oxide ceramics, so that the choice of matching metals is extremely limited. Therefore stress-relieving interlayers, such as ductile metal foils, are sometimes introduced between the ceramic and the metal [14].

Of course the interlayer metal should be able to interact with both the ceramic and the metal. As mentioned above, the formation of new phases at the ceramic–metal interface should be avoided as far as possible. Not only are the coefficients of thermal expansion of new phases usually widely different from those of the parent materials, but also the crystallographic structure can be completely different and this will lead to significant misfit stresses.

7.5 CONCLUSION

The conditions to be imposed on a ceramic–metal system for interfacial chemical reactions to proceed can be derived from equilibrium thermodynamic calculations. However, experiments have to be performed to determine the kinetics of the reactions. The formation of thick layers of intermetallics or non-metallic compounds should be avoided. Preference should be given to solid solutions since these have the same crystal structures as the parent material.

REFERENCES

1. Stoneham, A.M. and Tasker, P.W. (1985) *J. Phys. C*, **18**, 543.
2. Klomp, J.T. and van der Ven, A.J.C. (1980) *J. Mater. Sci.*, **15**, 2483.
3. Bueren, H.G. and van Hornstra, J. (1964) *Proc. 4th Int. Symp. on Reactivity of Solids*, Elsevier, Amsterdam.
4. Campbell, J.C. and Sherwood, E.M. (1967) *High Temperature Materials and Technology*, Wiley, New York.
5. *The Making, Shaping and Treating of Steel*, (1970), US Steel Corporation.
6. Barin, I. and Knacke, O. (1977) *Thermochemical Properties of Inorganic Substances*, Springer, New York.
7. Niessen, A.K., de Boer, F.R., Boom, R., de Chatel, P.F., Mattens, R.W. and Miedema, A.R. (1983) *Calphad*, **7**, 51.
8. Klomp, J.T. (1989) *Joining ceramics, glass and Metal* (ed. W. Kraft.) DGM Informations gesellschaft Verlag. Oberursel. pp 55–65.
9. Klomp, J.T. (1985) *Mater. Res. Soc. Proc.* **40**, 381.
10. Rhee, S.K. (1975) *J. Am. Ceram. Soc.*, **58**, 540.
11. Hansen, M. (1958) *Constitution of Binary Alloys*, McGraw Hill, New York.
12. Beyers, R., Sinclair, R. and Thomas, M.E. (1984) *J. Vac. Sci. Technol. B*, **2** (4), 781.
13. Creyke, W.E.C., Morell, R. and Sainsbury, I.E.J. (1982) *Design with Non-ductile Materials*, p. 206, Applied Science Publishers, London.
14. Nicholas, M.G. and Mortimer, D.A. (1985) *Mater. Sci. Technol.*, **1**, 657.

8

Bond strength of ceramic–metal joints

G. Elssner

8.1 INTRODUCTION AND GENERAL REMARKS

Bond strength measurements provide information on the mechanical quality and integrity of joints between metal and ceramic components. They can employ either conventional or fracture mechanics testing techniques.

In conventional bend, shear and tensile tests, the stress to fracture the bonded surfaces is used to characterize a bond strength. However, the failure strength of a joint manufactured from a brittle ceramic and a ductile metal is affected by stress concentrations at flaws present at the interface or in the adjacent ceramic. Typical flaws are cracks, pores and unbonded regions at interfaces or inclusions and cracks in the adjacent ceramic. Altering the distribution of flaw sizes by changing the dimensions of the bonded area therefore leads to a specimen size dependence for conventional bond strength values. Another important feature of conventional testing techniques arises from the fact that the bond strength can only be specified when it does not exceed the strength of the ceramics; otherwise fracture will occur away from the metal–ceramic interface region and the bulk of the brittle ceramic.

In fracture mechanics tests the interface fracture energy G_c is a measure of bond strength. Small sandwich bend test specimens are used which are notched by a sharp saw cut introduced into one of the interfaces between the central metal layer and adjacent ceramic pieces of the joint. The fracture energy is obtained from the failure load, the crack length (notch depth), the specimen dimensions and a correction function accounting for the geometry of the loaded specimen and the elastic properties of the constituents of the joint. In contrast with conventional bond strengths, the interface fracture energy is independent of the size of the interface area but it is influenced by the microstructure of the interface region. Interface fracture energy values can be used to derive the interface fracture resistance K_c of the joint. The advantage of the K_c measure of bond strength is its direct comparability with the fracture resistance K_{Ic} of monolithic materials.

8.1.1 Selecting a technique

The selection of an appropriate method for measuring the bond strength is dictated by the purpose of testing, but the bonding process and parameters affecting the mechanical quality of the bond can be monitored by both fracture mechanics and conventional testing methods. In the latter case, the dimensions of the specimens must be held constant because of size effects. The bond strength values obtained also depend on the testing technique chosen. Bend test values are generally higher than tensile test values for joints and for brittle ceramic materials.

Bond strength measurements are needed to monitor the development of new or improved bonding processes. Important parameters affecting the bond quality which have been or must be studied in this context, in addition to the usual bonding temperature, time and pressure, can be subdivided as follows:

1. mechanical properties of the constituents
2. chemical properties of the constituents
3. microstructure
4. properties and geometry of additionally introduced interlayers
5. surface characteristics such as roughness and contamination.

Values of the interface fracture energy G_c are required for fundamental studies of the influence of the structure and chemistry of metal–ceramic interfaces upon the mechanical behaviour of bonded systems. One way of obtaining these is to test metal–ceramic bicrystals with a well-defined mutual crystallographic orientation and known bulk and interface impurity contents.

In the following sections techniques for testing bond strength are outlined with emphasis on the fracture mechanics bond test method. Examples of bond strength measurements are given to illustrate the influence of bonding and material parameters on the mechanical quality of metal–ceramic joints.

8.2 TESTING METHODS

8.2.1 Conventional methods

Conventional testing methods can be classified into tensile, bend and shear tests. Typical tensile and bend test specimens are shown in Figs 8.1–8.4. The mechanical bond quality of brazed metal–ceramic joints is generally

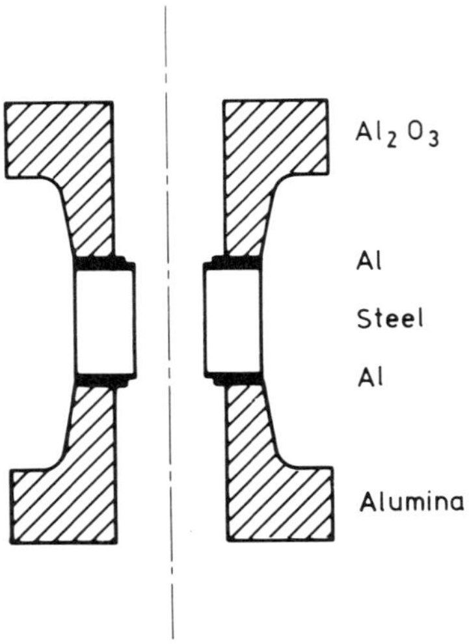

Fig. 8.1 ASTM brazed ceramic test pieces. The steel tubes are bonded to alumina (97.5%) with aluminium foil annuli.

assessed by the ASTM tensile test [1] or by shear tests. A detailed description of the advantages and disadvantages of shear tests has been given by Mizuhara *et al.* [2]. The shear strength is calculated by dividing the fracture load by the area of the brazed surface, but stresses at the joints

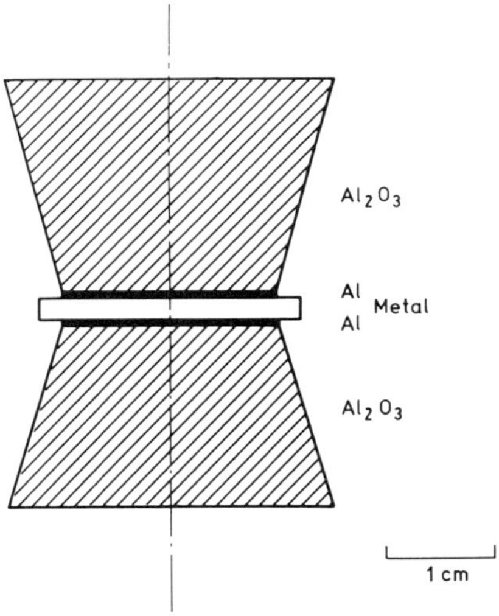

Fig. 8.2 Truncated cone bond strength specimens for tensile testing. 97.6% pure alumina pieces are bonded to metal discs with aluminium foils.

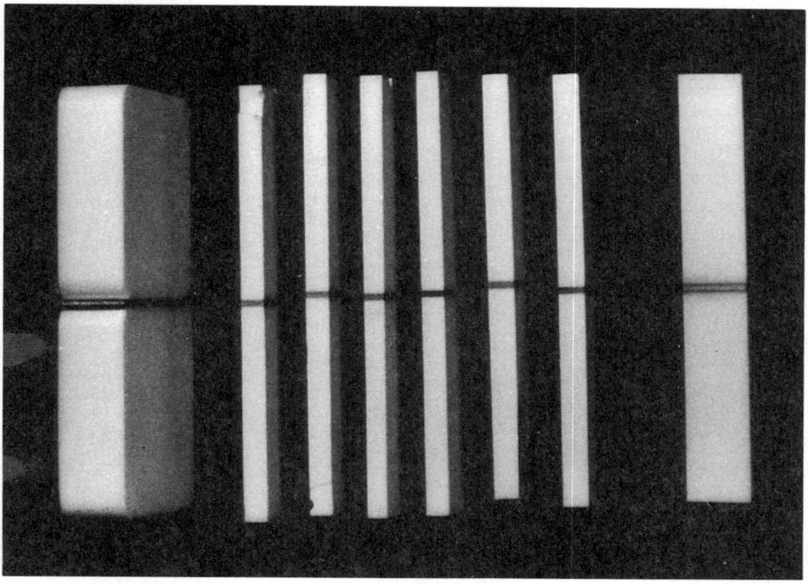

Fig. 8.3 Diffusion-bonded Al_2O_3–Nb–Al_2O_3 joint and bend test specimens cut from the joint (specimen length, 60 mm).

may not be pure shear because of the hold-down forces in ceramic block on metal block configurations or the thermal contraction mismatch in ceramic rod in metal ring configurations, both of which can affect the bond strength.

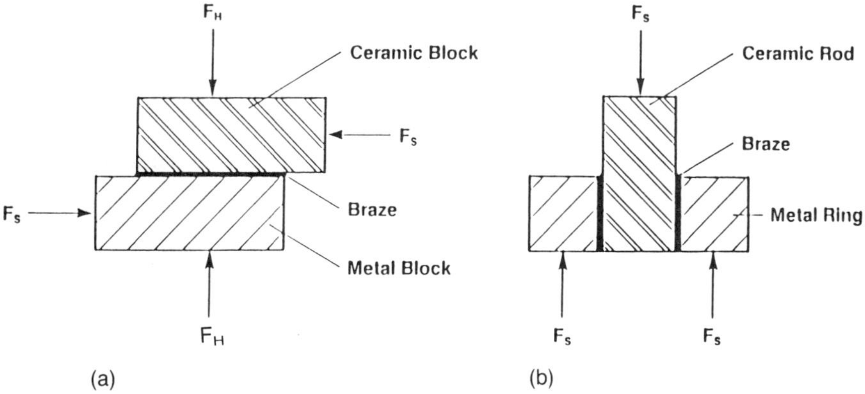

Fig. 8.4 Schematic illustrations of two common types of shear tests for brazed metal–ceramic joints: (a) ceramic block on metal block; (b) ceramic rod in metal ring.

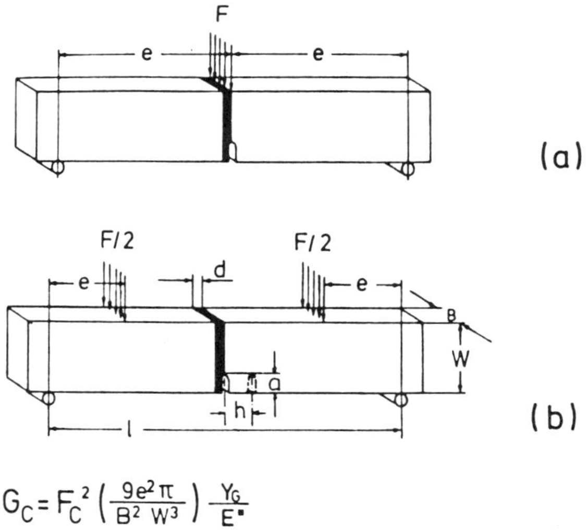

Fig. 8.5 Bonded bend test specimens and equation for the determination of the fracture energy G_c: (a) four-point bend test; (b) three-point bend test.

Three- and four-point bend tests (Fig. 8.5) have been used to determine the bond strengths of metal–ceramic joints. Up to now there are no national or international standards specifying loading conditions, shapes and dimensions of the specimens, but symmetrical specimens of rectangular cross-sections consisting of a metal layer in the middle and two adjacent ceramic parts are usually preferred. Typical dimensions of these sandwich specimens range from 30 mm × 5 mm × 2 mm to 60 mm × 7 mm × 4 mm with metal layer thicknesses between 10 μm and several millimetres. Other types of specimens used for bond strength determination include three-point specimens of circular cross-section composed of one metal and one ceramic part with the interface or additionally introduced interlayers in the middle. The flexural bond strength of sandwich specimens of rectangular cross-section can be determined from the fracture load F_c and the specimen dimensions (Fig. 8.5) from the equation

$$\sigma_{bf} = \frac{3F_c e}{BW^2} \tag{8.1}$$

8.2.2 Fracture mechanics bend test methods

In the fracture mechanics approach the mechanical quality of a joint is characterized principally by the fracture energy, but other useful measures are the fracture toughness and compliance.

The energy needed to separate the joint by a single crack propagating from a sharp interface notch or precrack into the ceramic–metal interface at the onset of catastrophic failure is the interface fracture energy G_c, which is also sometimes called the critical energy release rate or critical crack driving force. G_c, which has units of energy per unit area (J m^{-2}), is a material parameter affected only by the testing temperature and the deformation rate. It incorporates the reversible thermodynamic work of adhesion and the considerably larger irreversible contributions mainly originating from plastic deformation of the metal region adjacent to the running interface crack.

Fracture mechanics bend test techniques applied to metal–ceramic joints are similar to those used to determine the fracture toughness or resistance K_{Ic} of monolithic ceramics. However, in the latter case the evaluation of fracture mechanics strength parameters is simpler because the notch or the precrack is introduced into the middle of a bend test specimen made from a single material. Stresses and strains in the loaded specimen are therefore only affected by its loading geometry and by one set of elastic constants given by the modulus of elasticity E_c and the Poisson ratio μ_c of the ceramic material. The plane strain fracture toughness K_{Ic}, which is usually measured in units of MPa $m^{1/2}$, can be derived from the equation

$$K_{Ic} = \frac{3F_c e a^{1/2}}{BW^2} Y_k \tag{8.2}$$

where the fracture load is F_c and the correction function Y_k depends only on the loading mode for three- and four-point bend tests and on the ratio a/W of crack (notch) depth to specimen height (Fig. 8.5). Details of plane strain fracture toughness determination methods for ceramics and other monolithic materials can be found in the literature [3,4].

Similarly, mechanical quality can be defined in terms of a fracture energy G_c which for a monolithic ceramic bend strength sample can be derived from the expression

$$G_c = \frac{pa}{E_c} (1 - \mathrm{m}_c^2)\, \mathrm{d}_{bf}^2 \tag{8.3}$$

where E_c is Young's modulus of elasticity and m_c is Poisson's ratio for the ceramic.

Rigorous determination of G_c values for ceramic–metal couples is more difficult because of, for example, uncertainties about what value to use for the elastic modulus term. Nevertheless the concepts of both fracture toughness and fracture energy are useful, and values are being increasingly quoted in the literature. The analytical interpretation of fracture mechanics bend test methods are discussed further in the Appendix.

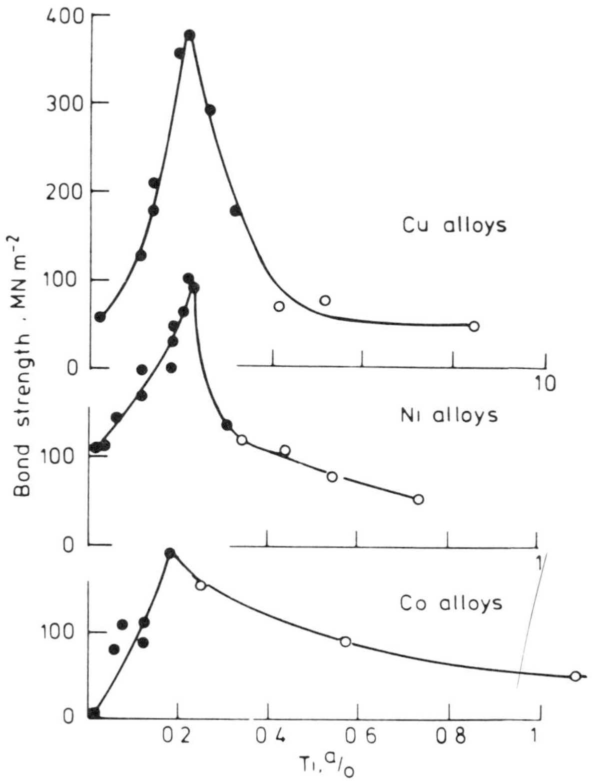

Fig. 8.6 The tensile strength of alumina interfaces formed with some titanium-containing alloys by active brazing. The open symbols refer to wetted interfaces.

8.3 BOND STRENGTH MEASUREMENTS

8.3.1 Conventional methods

Conventional methods of bond strength measurement have been used to optimize the composition of brazing alloys for metal–ceramic joints. Thus tensile strength data obtained from push-off tests of solidified sessile drops on alumina plates (Fig. 8.6) have shown that alumina interfaces formed with copper, nickel and cobalt alloys exhibit a maximum strength at certain contents of the active metal titanium [5]. Furthermore, it has been shown that for given base alloys, e.g. nickel alloys brazed to alumina [6], the tensile strength values for interfaces possessing optimized contents of the active metals chromium, vanadium, titanium, aluminium, thorium and yttrium increase linearly with the absolute values of the free energy of oxide formation of the active metals (Fig. 8.7). These correlations between

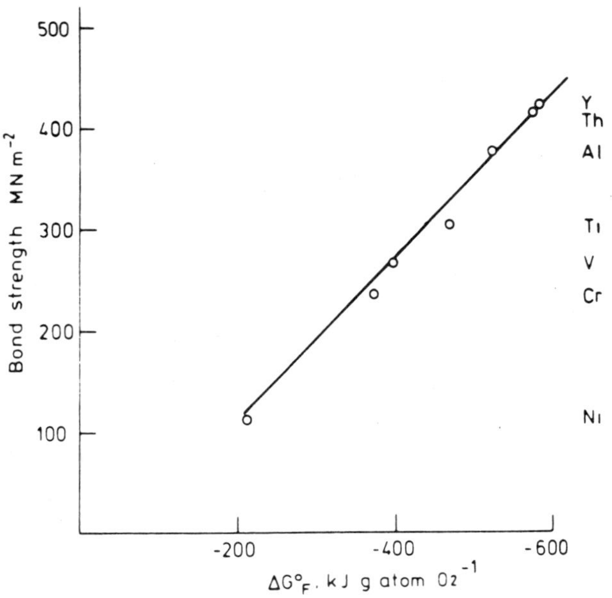

Fig. 8.7 Correlation between the chemical reactivity of alloying additions and the maximum strengths of brazed bonds formed between alumina (99.5%) and nickel-base alloys. The tensile strength values were calculated from push-off test data of solidified metal drops on ceramic plates.

chemical reactivity and bonding behaviour could be detected because the mechanical test configurations were kept constant and only minor changes in the mechanical bulk properties of the metal component of the joints were introduced by the additions of active metal.

Differences in the thermal contraction behaviour of the constituents can strongly affect the bond strength of joints. For example, in a study on diffusion bonding of metals to alumina using aluminium interlayers it was observed that the tensile strengths obtained using truncated cone specimens (Fig. 8.2) decreased with increasing thermal expansion coefficient of the 3 mm thick metal discs inserted (Fig. 8.8). The thermal expansion coefficient of the alumina ceramic was $7 \times 10^{-6}\ K^{-1}$ and the bonding temperature was 550 °C. The marked thermal expansion mismatch of the copper imposed residual stresses in the interface region which, together with the formation of brittle and flawed intermetallic layers between copper and aluminium, resulted in minimum values of bond strength [7].

The interfacial tensile strength of samples manufactured from Si_3N_4 and steel discs shows a strong dependence on the thickness of the ceramic component [8]. For steel discs of 5 mm diameter bonded to ceramic by hot isostatic pressing at 1200 °C, a decrease in strength from about 50 MPa to

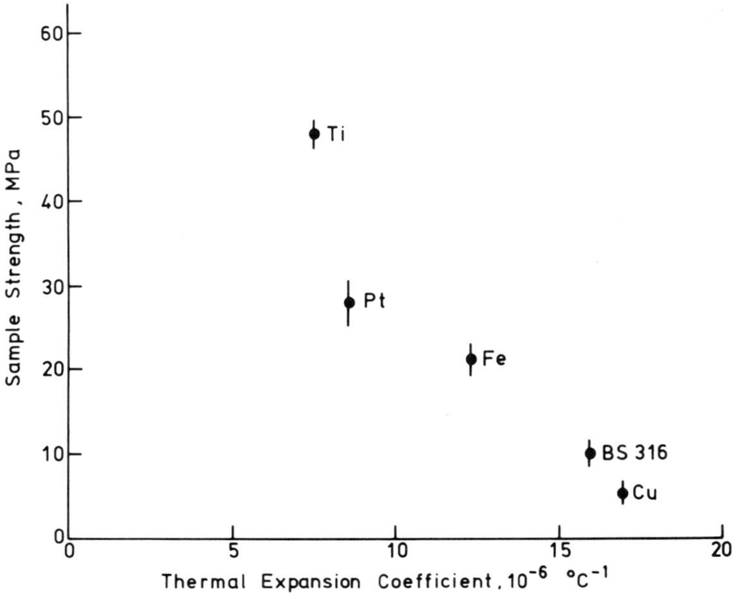

Fig. 8.8 Tensile strengths of truncated cone alumina–metal samples bonded with aluminium foil plotted as a function of the thermal expansion coefficients of the metal (BS 316 is CrNiTi stainless steel).

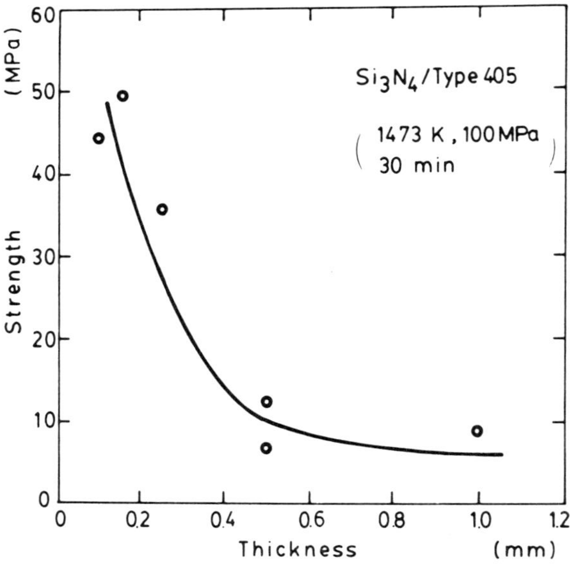

Fig. 8.9 Thickness dependence of the tensile strength of an Si_3N_4–steel joint diffusion bonded at 1200 °C for 30 min under a pressure of 100 MPa.

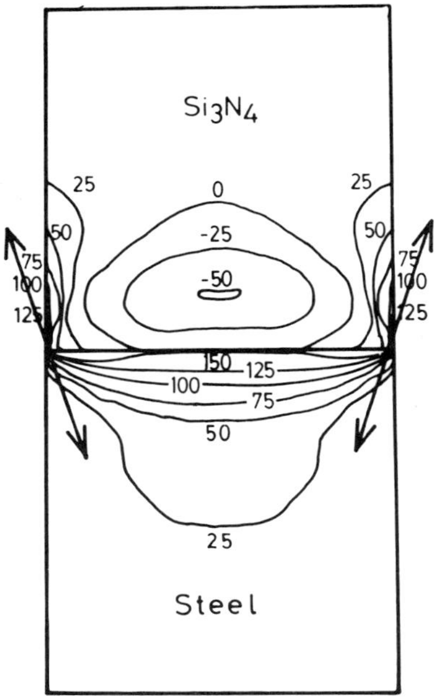

Fig. 8.10 Contour map of the internal stress in an Si_3N_4–steel joint calculated using the finite-element method assuming fully elastic materials and a difference of 975 K between the bonding temperature and room temperature. The arrows represent the maximum tensile stress appearing in the silicon nitride. The numbers on the contours must be multiplied by 10 to give the stresses in megapascals.

10 MPa with increasing thickness of the ceramic from 0.1 to 10 mm was observed (Fig. 8.9). This effect is mainly due to internal stresses caused by the large mismatch in the contraction behaviour of the materials; the thermal expansion coefficients of steel and Si_3N_4 are $14 \times 10^{-6}\ K^{-1}$ and $2.5 \times 10^{-6}\ K^{-1}$ respectively. A contour map of internal stresses in an Si_3N_4–steel joint calculated using the finite-element method assuming fully elastic materials and a difference of 975 K between bonding and room temperature is shown in Fig. 8.10. According to this map, the maximum tensile stresses will be at the edges of the interface within the ceramic as suggested by the fracture appearance of the joints [1]. The deformation induced by the thermal contraction mismatch of an Si_3N_4–steel joint cooled from a bonding temperature of 1000 °C to room temperature is illustrated in Fig. 8.11. Furthermore, finite-element calculations show that the maximum residual stress in the ceramic will increase with increasing thickness of the Si_3N_4 layer (Fig. 8.12), which explains the observed drastic reduction in conventional bond strength.

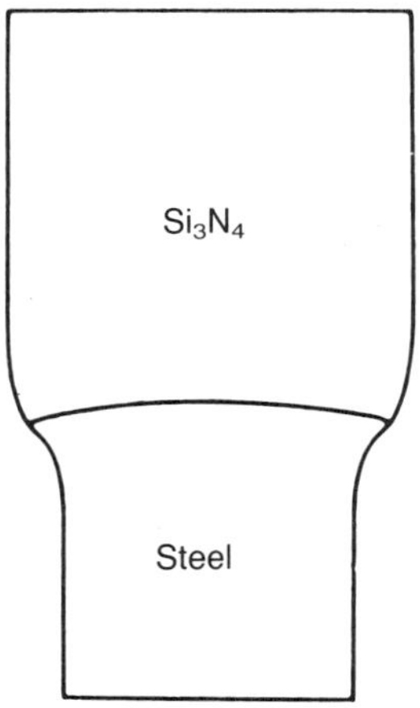

Fig. 8.11 Strain in an Si_3N_4–steel joint cooled from a bonding temperature of 1000 °C to room temperature. The deformation is magnified by a factor of 25.

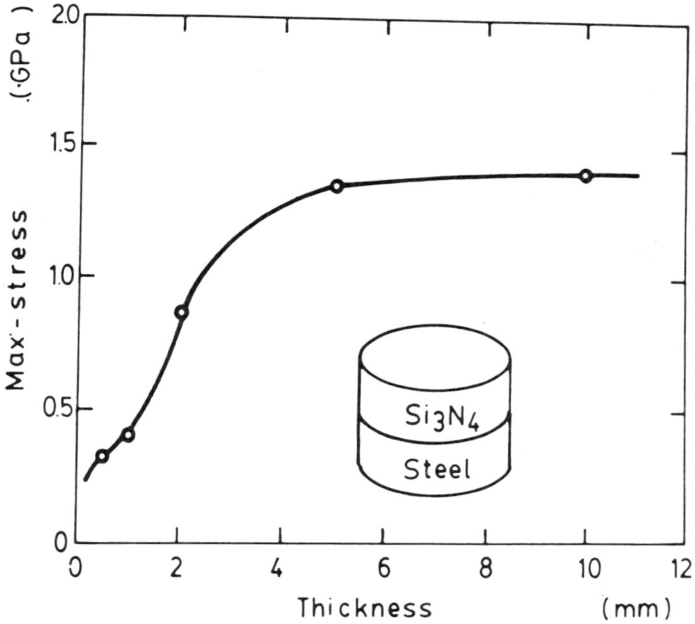

Fig. 8.12 Calculated maximum residual tensile stress in the Si_3N_4 component of an Si_3N_4–steel joint as a function of the thickness of Si_3N_4. The thickness of the steel disc is 10 mm, and the specimens are assumed to be cooled from 1000 to 25 °C.

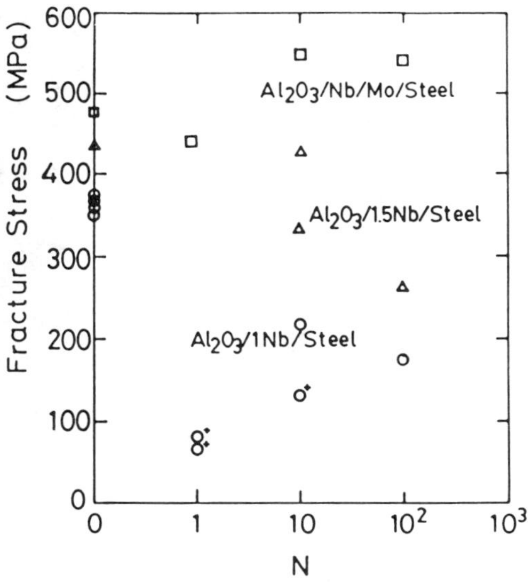

Fig. 8.13 Fracture stress of four-point bend test specimens (2 mm × 2 mm × 15 mm) cut from larger (5 mm diameter) Al_2O_3–steel pieces with Nb–Mo (0.5 mm and 0.5 mm) and niobium (1.5 mm and 1 mm) interlayers as a function of the number *N* of thermal cycles between room temperature and 500 °C. For comparison, an original bonded Al_2O_3–Nb–Mo–Nb–steel piece of diameter 5 mm failed at a tensile stress of 63 MPa at the Al_2O_3–Nb interface.

Similar tendencies are found for the dependence of the bond strength on the size of the bonded surfaces. In Fig. 8.13 the fracture stress of four-point bend test specimens with 2 mm × 2 mm cross-sections cut from alumina–steel pieces with Nb/Mo and niobium interlayers is plotted against the number of thermal cycles between room temperature and 500 °C. It is interesting to note that the measured interface bond strength (350–380 MPa) of the as-cut bend test specimens with a niobium interlayer 1 mm thick was appreciably larger than the highest observed tensile interface fracture stress (63 MPa) of an original diffusion-bonded piece with an Nb/Mo double interlayer [8]. However, thermal cycling of the bend test specimens reintroduced residual stresses and caused a reduction of the fracture stress.

8.3.2 Fracture mechanic bend test methods

Turwitt *et al.* [9] used polycrystalline pieces of 99.7 wt% Al_2O_3 and 99.8 wt% niobium sheets 1 mm thick to study the influence of surface preparation on the bond strength of diffusion-bonded metal–ceramic joints. The Nb–Al_2O_3 system was chosen because there is a minimal thermal misfit

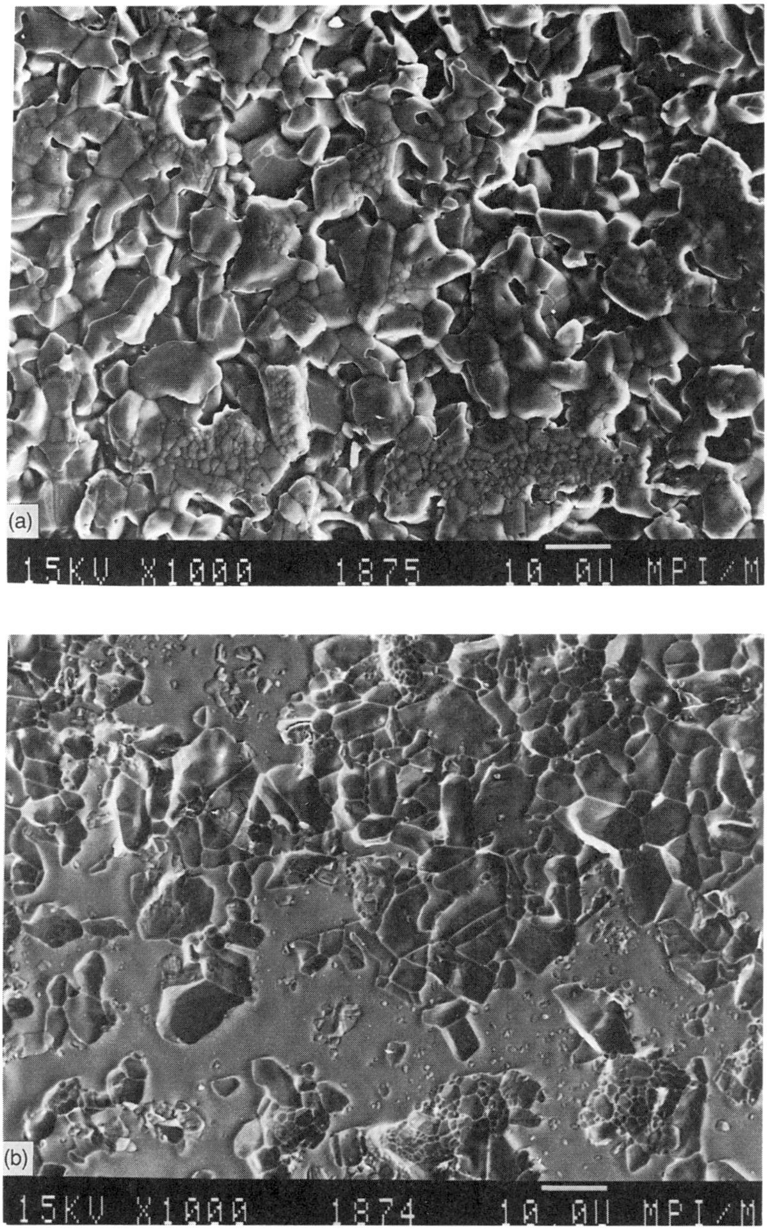

Fig. 8.14 Scanning electron micrographs of the fracture surfaces of a niobium–alumina joint manufactured by diffusion bonding at 1700 °C for 2 h at 10 MPa from as-cut alumina pieces and as-polished niobium sheet. Fracture at the interface: (a) alumina side and (b) niobium side.

and no intermediate reaction layers are formed at the metal–ceramic interface in oxygen-free environments.

Interfaces with a different degree of bonding were produced using as-cut, as-ground and as-polished (1 μm diamond paste) alumina bodies together with polished niobium sheets. Two different final grinding operations were applied using diamond discs of grit size 64 μm and 20 μm. Sandwich Al_2O_3–Nb–Al_2O_3 joints were manufactured at 1700 °C for 2 h at 10 MPa in a high-vacuum chamber. Six four-point bend test specimens (Fig. 8.3) were cut from one bonded piece and a notch 150 μm wide was introduced into one interface (Fig. 8.5). Scanning electron micrographs of the fracture surfaces of a joint made from as-cut alumina pieces are shown in Fig. 8.14. Fracture always occurred at the interface. Non-bonded areas are visible as smooth regions on the metal side of the fracture surface (Fig. 8.14(b)). However, a scanning electron micrograph of the niobium side of the fracture surface of a joint fabricated from as-polished alumina pieces (Fig. 8.15) did not reveal any unbonded areas. The fracture surface of the most carefully bonded joint displayed only the imprint of the interface grain structure of the ceramic.

In this study the ratio $P_H = A_{ap}/A_p$, where A_{ap} is the projected bonding interface area and A_p is the projected total interface area, was taken as a

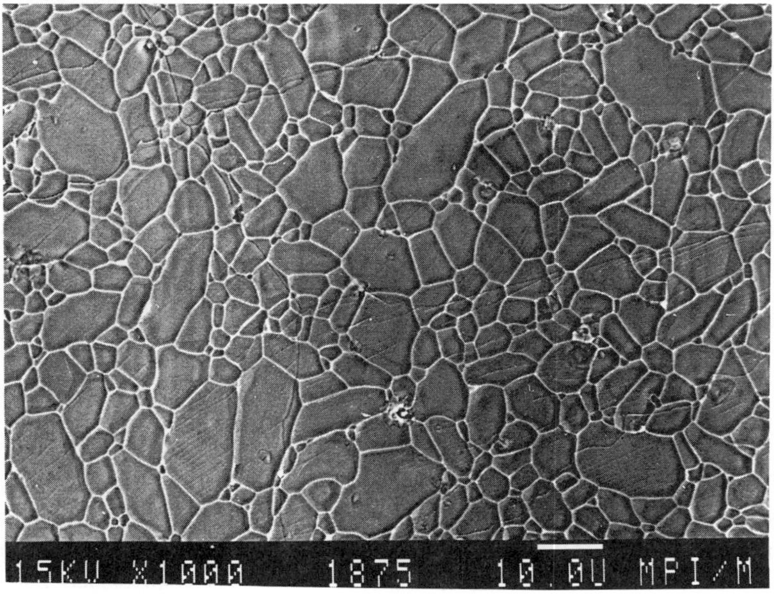

Fig. 8.15 Scanning electron micrograph of the niobium side of a fracture surface taken from a niobium–alumina joint manufactured by diffusion bonding at 1700 °C for 2 h at 10 MPa from as-polished alumina pieces and as-polished niobium sheet. Fracture at the interface.

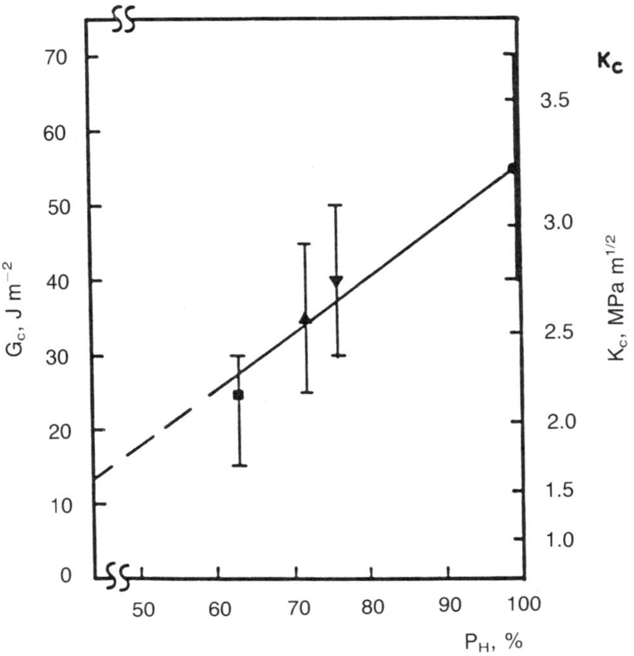

Fig. 8.16 Interface fracture energy G_c as a function of the degree of bonding P_H.

measure of the degree of bonding. The degree of bonding was measured at the fracture surfaces by scanning electron microscopy using a point-count method [9]. Figure 8.16 shows that the interface fracture energy G_c increases with increasing P_H and that the joints manufactured from polished ceramic and metal pieces are characterized by a maximum value of fracture energy and a P_H value of about 100%. Slight deviations from ideal bonding can be related to the presence of pores at the alumina interface. This influence of surface preparation corresponds to results of similar studies performed with other bonded systems fabricated by both diffusion bonding and active brazing.

Measurements of the influence of metal grain size on the interface fracture energy have also been obtained using polished sapphire (Al_2O_3 single crystal) plates with a $(1\bar{1}00)$ orientation and a thickness of 1 mm sandwiched between two polished niobium plates 1 mm thick of different initial grain size and diffusion bonded at 1700 °C for 2 h at 10 MPa in a high vacuum [10]. The configuration of a four-point bend test specimen cut from bonded pieces is shown in Fig. 8.17. The measured interface fracture energy is plotted in Fig. 8.18 against the difference $\bar{D}_a - \bar{D}_b$ between the niobium grain sizes after and before bonding, and it can be seen that G_c increases with increasing grain growth of the metal. In explaining these results it can be argued that extensive grain growth in the metal component

Fig. 8.17 Four-point bend test specimen of configuration alumina–(niobium single crystal)–sapphire–(niobium single crystal)–alumina notched at one interface between sapphire and niobium. Thickness of sapphire platelet, 1 mm.

during diffusion bonding is accompanied by an accommodation of interface structure and orientation between each grain and the adjacent plane of the sapphire crystal, yielding a low-energy configuration with a high interface fracture energy.

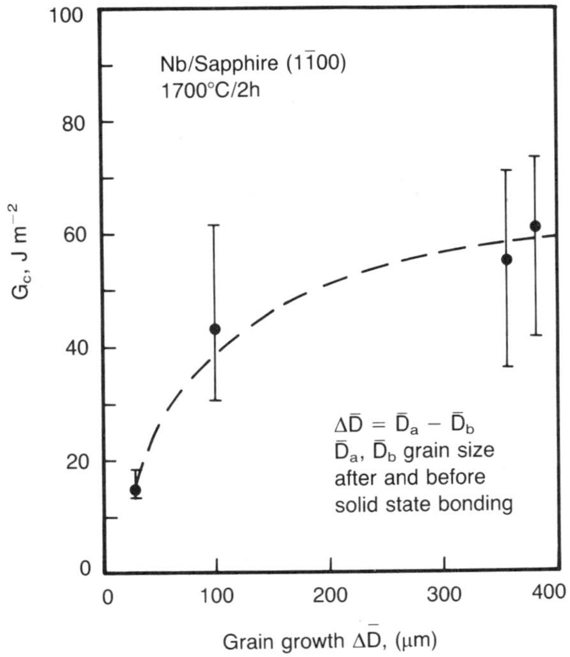

Fig. 8.18 Interface fracture energy G_c of joints between (1$\bar{1}$00) sapphire and polycrystalline niobium as a function of the niobium grain growth $\Delta\bar{D}$.

Table 8.1 Values of the interface fracture energy G_c for niobium–sapphire joints [11]

	Niobium (polycrystalline), $D = 370\ \mu m$	*Niobium (single crystal, (110) plane)*
Sapphire (0001)	$G_c = 26\ J\ m^{-2}$	$G_c > 150\ J\ m^{-2}$ $[1\bar{1}0]_{Nb} \parallel [11\bar{2}0]_{sapphire}$
Sapphire (1$\bar{1}$00)	$G_c = 55\ J\ m^{-2}$	$G_c = 74\ J\ m^{-2}$ $[001]_{Nb} \parallel [0001]_{sapphire}$
		$G_c = 110\ J\ m^{-2}$ $[001]_{Nb} \parallel [11\bar{2}0]_{sapphire}$

The assumption of an orientation dependence of interface fracture energy is based on experimental results [11] obtained from both single-crystal couples and single-crystal–polycrystalline Nb–Al_2O_3 couples diffusion bonded at 1700 °C for 2 h using a pressure of 10 MPa in a high vacuum. Four-point bend test specimens (Fig. 8.17) were used to determine the values of the interface fracture energy G_c given in Table 8.1. The results indicate that G_c values are influenced by the type of crystallographic planes mating at the interface and, for a given pair of mating planes, by the directions aligned parallel to each other in the niobium and sapphire planes. The observed dependence may be caused by orientation-dependent energy dissipation processes during fracture, by differences in the interface structure or by a combination of both.

The final example relates to the interface strength of Cu–Al_2O_3 joints diffusion bonded in an ultrahigh vaccum. G_c values for notched four-point bend test specimens [12] are used to assess the effects of bonding temperature on the quality of the metal–ceramic joints. The surfaces of polished 99.99 wt% copper plates (2 ppm oxygen) and polished alumina pieces (99.7 wt% Al_2O_3) were sputter cleaned by bombardment with 3 kV argon ions. The cleaned samples were stacked in a vacuum of 1×10^{-10} mbar in an Al_2O_3–Cu–Al_2O_3 configuration and bonded at various temperatures for 2 h. During heating, the vacuum pressure increased to about 2×10^{-8} mbar. The results (Fig. 8.19) demonstrate that strong bonds between copper and alumina can be obtained by ultrahigh vacuum bonding of sputter-cleaned surfaces, and transmission electron microscopy observations showed that no intermediate spinel layer was formed. For each bonding temperature, 10–20 G_c measurements were performed at room temperature. The increase in interface fracture energy with increasing bonding temperature is due to an increase in intimate contact between the surface of the metal and the ceramic; considerable surface shape

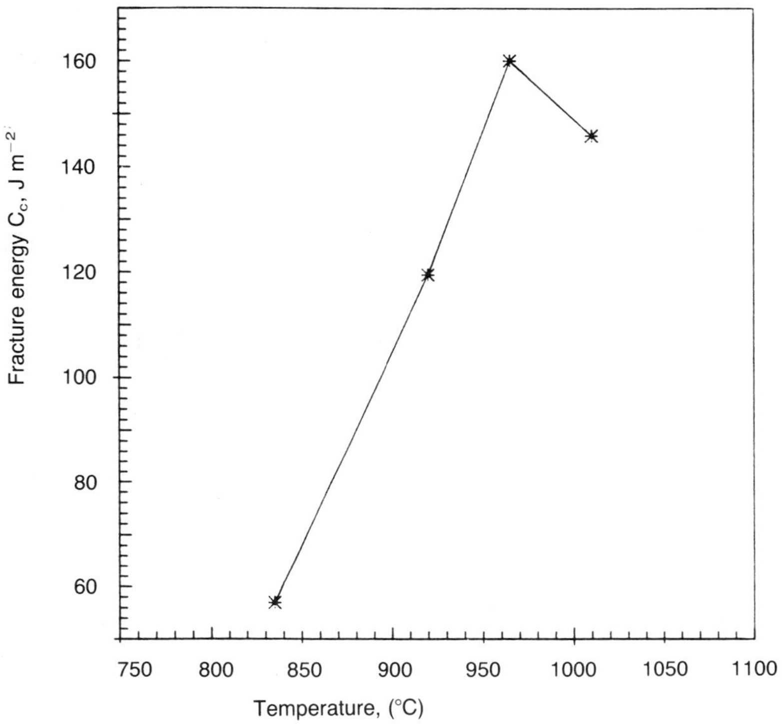

Fig. 8.19 G_c values for interfaces of polycrystalline Al_2O_3 against polycrystalline copper produced in an ultrahigh vacuum by bonding sputter-cleaned surfaces under a pressure of 16 MPa at the temperatures shown [12].

modification is necessary to create a perfect pore-free bond between such surfaces.

Figure 8.20 shows the interface fracture resistance K_c of some bonded systems in comparison with the fracture toughness data K_{Ic} of a number of engineering materials [3]. The K_c data were calculated from measured G_c values.

8.4 CONCLUDING REMARKS

Some warnings are necessary with respect to the often stated requirement that bond strength tests should be meaningful, reproducible, unambiguous and related to the reliability of the component or device in which the metal–ceramic joint is incorporated. The measured bond strength values will be meaningful and reproducible when precautions concerning the size and testing-mode dependence of conventional data are taken into account. An assessment of the mechanical quality of the joints in production control

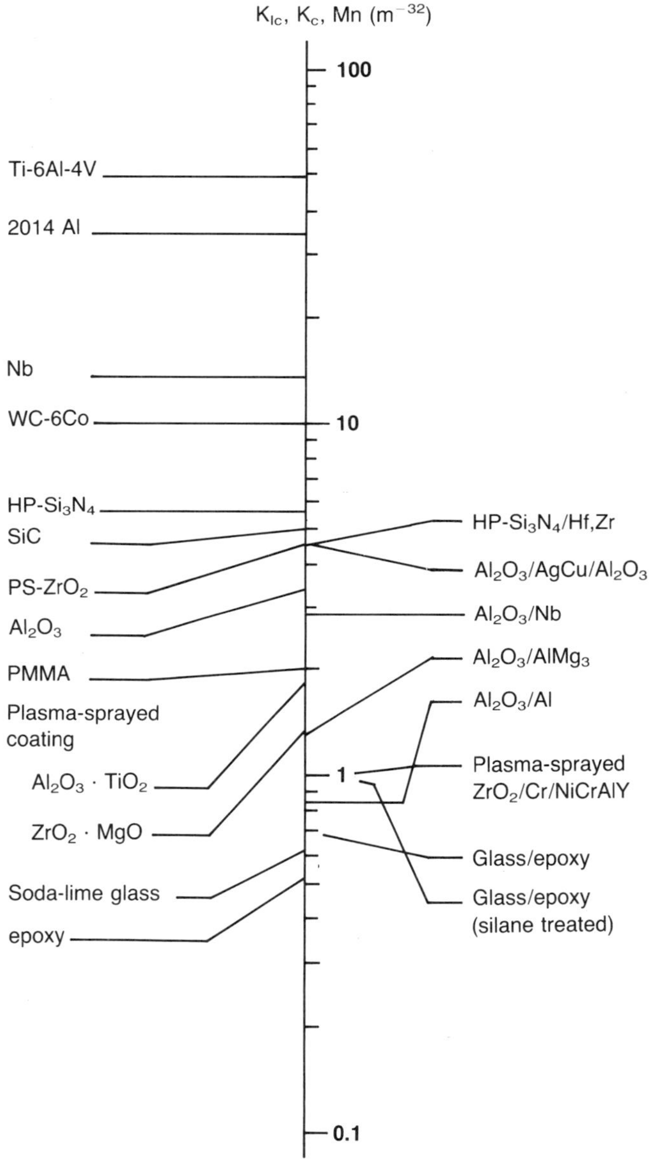

Fig. 8.20 Fracture resistance data K_{Ic} for some engineering materials and K_c for typical material joints.

and research and development is thereby guaranteed. However, total reliance on strength measurements to characterize the mechanical behaviour of bonded systems in service conditions is dangerous. The failure of a mechanically loaded ceramic–metal component is strongly influenced by both the severity of the flaws present and the local variation of stresses governed by the geometry of the joint, by the elastic and plastic properties of the constituents and by the loading conditions of the bonded system. Furthermore, the influence of residual stresses due to differences in the thermal contraction and expansion behaviour of the constituents of the system cannot be neglected. Accordingly, the quality of metal–ceramic joints must be tested under the actual service conditions and assembly configurations to ensure their reliability.

Test specimens are often cut from larger bonded metal–ceramic pieces and must be shaped by grinding and polishing. The extent and the distribution of residual stresses in the specimens can be reduced by such operations, and unrealistically high values of conventional bond strength may be obtained with 'cut-out' specimens if the thermal contraction mismatch of the constituents of the joint is large. However, the possible degradation of the joint properties can easily be revealed by thermal cycling of the cut-out specimens and subsequent determination of the bond strength [7].

Specimens cut from larger bonded pieces can be used to monitor the uniformity of the bond quality along the interface region. Unbonded regions and other defects lead to a decrease of both conventional and fracture mechanics bond strength values. Thus it must be emphasized that bond strength evaluation should be accompanied by thorough optical and electron microscopy examinations of cross-sections through the interface region and scanning electron microscopy observations of the fractured surface to obtain a deeper insight into the interrelation between defect size distribution, microstructure and mechanical behaviour.

APPENDIX: ANALYTICAL INTERPRETATION OF FRACTURE MECHANICS BEND TESTS

The plane strain fracture toughness of a monolithic ceramic bend test sample can be derived from the relationship

$$K_{\mathrm{Ic}} = \frac{3F_{\mathrm{c}}\, e\, a^{1/2}}{BW^2} Y_{\mathrm{K}}$$

where F_c is the failure load, a is the notch depth, Y_K is a correction factor identified later, and e, B and W are sample dimensions identified in Fig. 8.5.

Plane strain fracture energies G_{Ic} can be calculated from K_{Ic} data for ceramics using the equation

$$G_{Ic} = (1 - \mu_c^2)\,\frac{K_{Ic}^2}{E_c} \qquad (8A.1)$$

if the modulus of elasticity E_c and the Poisson ratio μ_c are known. (Eqn (8A.1) can also be derived from eqn (8A.8) given below for joints made from dissimilar materials by setting $\beta' = 0$, $E^* = E_c/(1 - \mu_c^2)$, $K_c = K_{Ic}$ and $G_c = G_{Ic}$.)

Generally, fracture energies are derived by using a total energy balance for the creation of new surfaces. If kinetic energy effects, heat fluxes and body forces can be neglected, as is usually the case in laboratory tests, the interface fracture energy G_c will be given by

$$G_c = \frac{d(U_E - W_s - W_D)}{dA} \qquad (8A.2)$$

where A is the crack surface, U_E is the internal energy, W_s is the work due to pure surface creation and W_D is the dissipated work.

The unequal elastic properties of the two bonded materials will introduce an element of plane shear (mode II) crack opening even when the external forces are applied in tension to cause opening (mode I) plane strain. Therefore the tensile stress s_y and the shear stress t_{xy} at the crack tip perpendicular to or within the crack plane can be described by an equation of the following type [13]:

$$(\sigma_y - \mathrm{i}\tau_{xy})_{\theta=0} = \frac{(K_I - \mathrm{i}K_{II})}{(2\pi r)^{1/2}} \qquad (8A.3)$$

where K_I and K_{II} are the mode I and mode II stress intensity factors, $\mathrm{i} = (-1)^{1/2}$, r is the radius and $\theta = 0$ is the angle made by the radius r with the interface ahead of the crack in a two-dimensional model of the crack-tip region. The absolute value of the complex stress intensity factor K of an interfacial crack is independent of the assumptions of the crack-tip model [13] and is given by

$$K = (K_I^2 + K_{II}^2)^{1/2} \qquad (8A.4)$$

It is important to note that a unique relationship holds between K and the energy release rate G:

$$G = \frac{(1 - \beta'^2)K^2}{E^*} \qquad (8A.5)$$

where E^* is an effective modulus of elasticity

$$\frac{1}{E^*} = \frac{1}{16}\left(\frac{1+\kappa_c}{S_c} + \frac{1+\kappa_m}{S_m}\right) \tag{8A.6}$$

and β' is the Dundurs parameter [14]

$$\beta' = \frac{k(\kappa_c - 1) - (\kappa_m - 1)}{k(\kappa_c + 1) + (\kappa_m + 1)} \tag{8A.7}$$

S^* is the shear modulus, $k = S_c^*/S_m^*$ and k = (3 − m)/(1 + m) for plane stress or k = 3 − 4m for generalized plane strain where m is Poisson's ratio.

According to eqn (8A.5) the fracture resistance K_c can be derived from the fracture energy G_c:

$$K_c = \left(\frac{E^* G_c}{1 - \beta'^2}\right)^{1/2} \tag{8A.8}$$

Therefore K_c is an alternative measure of the bond strength and is usually given in units of MPa $m^{1/2}$.

The interface fracture energy G_c of metal–ceramic joints can be determined from three- and four-point bend tests. If the load–deflection curve is linear up to the fracture load F_c the fracture energy can be calculated from the experimental data using the equation

$$G_c = F_c^2 \frac{9e^2\pi}{B^2W^3}\frac{Y_G}{E^*} \tag{8A.9}$$

where e is the difference between the outer and inner span of the four-point bend test device, B is the width and W the height of the specimen, and Y_G is a correction function (Fig. 8.5). Otherwise, a non-linear coefficient C' must be employed to obtain a generalized fracture energy $\bar{G}_c = C'G_c$. The non-linearity coefficient is determined experimentally from the load–displacement curve of the specimen [15].

Typical specimen lengths are 30–60 mm, and the height W and width B range from 6 to 10 mm and 2 to 10 mm respectively. The correction function Y_G depends on the ratios a/W, d/W and h/W and on the Dundurs parameters a' and β', where a is the crack length or notch depth, d is the thickness of the metal layer and h is the distance between the notch position and the metal–ceramic interface.

The Dundurs parameter a' can be derived from

$$a' = \frac{K(\kappa_c + 1) - (\kappa_m + 1)}{K(\kappa_c + 1) + (\kappa_m + 1)} \tag{8A.10}$$

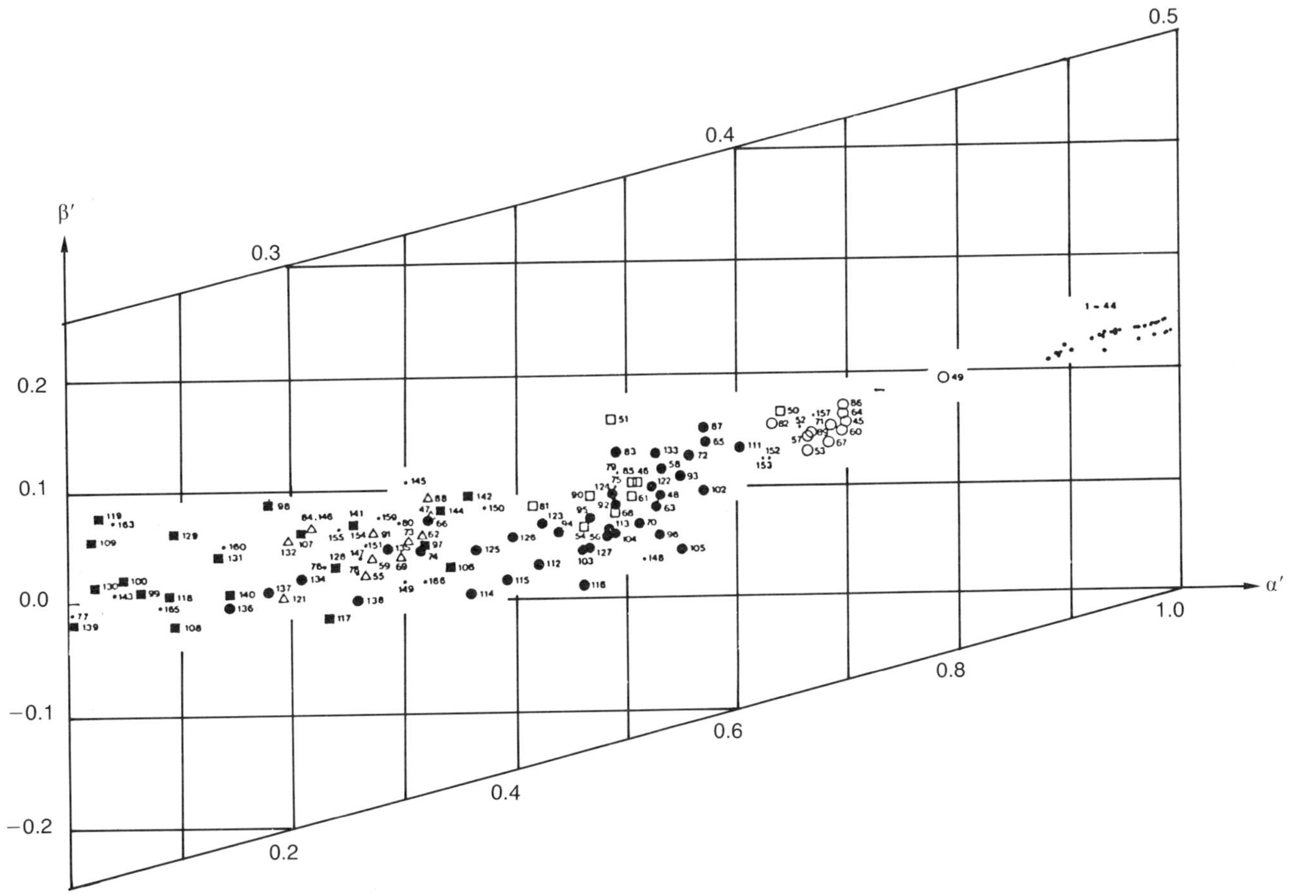

Fig. 8.21 Correlation between the Dundurs composite parameters a′ and b′ for some material combinations: ○, Al/ceramic, Mo, W, B, C; □, Cu/ceramic, Mo, W, B, C; △, Ni/ceramic, Mo, W, B, C; ●, Ti, Zr, Hf, V, Nb/ceramic, Mo, W; ■, Ta, Cr, Mo, W/ceramic; ·, 1–44, plastics; ·, others. The numbers denote the material combinations.

and Fig. 8.21 shows the correleation of α' and β' for some material combinations [16]. If the crack extension is uniform across the specimen width B, the energy release rate of a specimen can be evaluated from the applied force F and the derivative of the compliance $C' = \delta/F(\delta$ is the displacement) with respect to the crack length a:

$$G = \frac{F^2}{2B}\frac{\mathrm{d}C'}{\mathrm{d}a} \tag{8A.11}$$

where the compliance C' of notched bend test specimens is given by

$$C' = C'_0 + \frac{18e^2\pi}{BW^2E^*}\int Y_G \,\mathrm{d}\left(\frac{a}{W}\right) \tag{8A.12}$$

and C'_0 is the value of the compliance of a specimen without a crack or notch. The introduction of a normalized compliance

$$C^* = (C' - C'_0)\frac{BW^2E^*}{18e^2\pi} \tag{8A.13}$$

leads to the relation

$$Y_G = \frac{\partial C^*}{\partial(a/W)} \tag{8A.14}$$

for the correction function Y_G, according to which Y_G is the derivative of the normalized compliance with respect to the normalized crack length a/W. This relation is the basis of both the experimental determination and the finite-element calculation of the correction function Y_G. An example of calculated and experimentally determined correction functions is given in Fig. 8.22, in which the C^* and Y_G values of HPSN–Hf–HPSN (HPSN, hot-pressed Si_3N_4) are plotted against the ratio a/W. The values of the normalized compliance were calculated using the virtual crack extension method. The curves of the correction functions for various d/W ratios are bounded by the Y_G curve ($d/W = \infty$) which is closest to the Y_{iso} curve for a homogeneous single material and by the curve $Y_0 = (1 + \alpha)Y_{iso}$ for $d/W = 0$ with $\alpha' = 0.3664$.

The correction function can also be estimated by an approximation method [16] based on the following equations:

$$Y_G = \frac{Y_{bi} - Y_0}{1 + \exp\{-f(x)\}} + Y_0 \tag{8A.15}$$

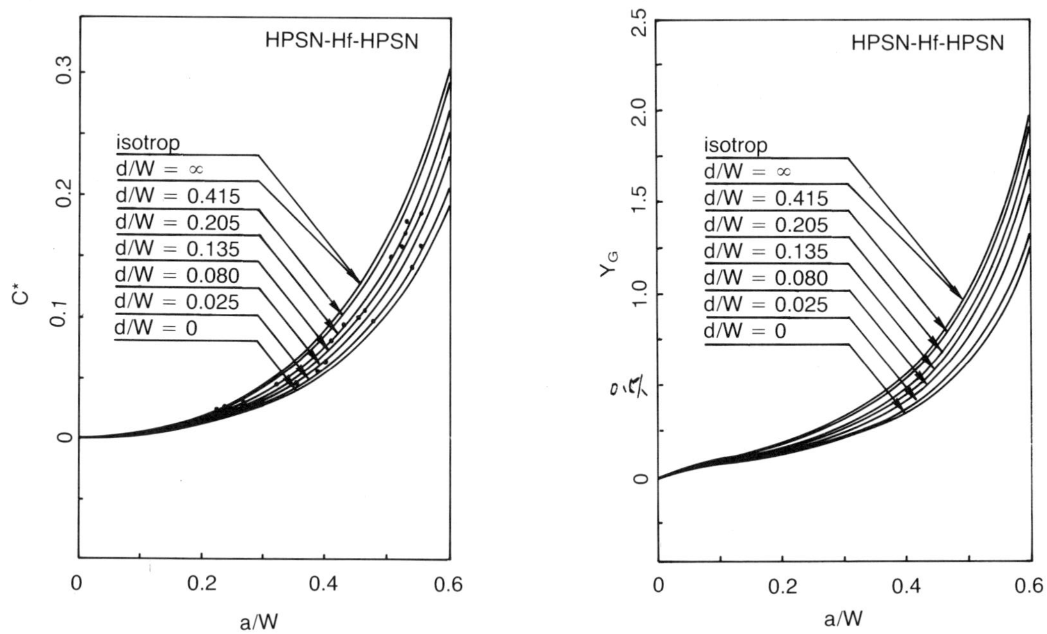

Fig. 8.22 Normalized compliance C^* and correction function Y_G plotted against the ratio a/W for HPSN–Hf–HPSN joints (three-point bend test): ·, measured values; ——, calculated values.

$$Y_{bi} = (1 - \beta'^2)(1 + 4\varepsilon^2)Y_{iso} \qquad (8A.16)$$

$$e = \frac{1}{2}\ln\left(\frac{1 + b'}{1 - b'}\right) \qquad (8A.17)$$

$$Y_0 = 1(1 + a')Y_{iso} \qquad (8A.18)$$

$$x = \ln\left(\frac{d}{W}\right) \qquad f(x) = 4.2709 + 2.8063x + 0.2921x^2 \qquad (8A.19)$$

where e is a bimaterial constant. Y_{iso} can be calculated from the correction function Y_K used for the determination of fracture toughness K_{Ic} in three- and four-point bend testing of notched specimens:

$$Y_{iso} = \frac{1}{p}\frac{a}{W}Y_K^2 \qquad (8A.20)$$

For a three-point bend test [17]

$$Y_k = 1.96 - 2.75\frac{a}{W} + 13.66\left(\frac{a}{W}\right)^2 - 23.98\left(\frac{a}{W}\right)^3 + 25.22\left(\frac{a}{W}\right)^4 \qquad (8A.21)$$

and for a four-point bend test [17]

$$Y_k = 1.96 - 2.47\frac{a}{W} + 12.97\left(\frac{a}{W}\right)^2 - 23.17\left(\frac{a}{W}\right)^3 + 24.80\left(\frac{a}{W}\right)^4 \qquad (8A.22)$$

The application of the approximation method to specimens $a/W = 0.3$ and an interface crack leads to Y_G data with a 10% maximum deviation from those calculated using finite-element methods.

REFERENCES

1. *Standard Method for Tension and Vacuum Testing Metallized Ceramic Seals*, ASTM F19–64, 1964.
2. Mizuhara, H., Huebel, E. and Oyama, T. (1987) *Am. Ceram. Soc. Symp. on Structural Ceramic Joining*, in press.
3. Broek, D. (1974) *Elementary Engineering Fracture Mechanics*, Noordhoff.
4. Munz, D., Bubsey, R.T. and Shannon, J.L., Jr. (1980) *J. Am. Ceram. Soc.*, **63**, 300.
5. Nicholas, M.G. (1969) *Sci. Ceram.*, **6**, 214.
6. Crispin, R.M. and Nicholas, M.G. (1976) *J. Mater. Sci.*, **11**, 17.
7. Nicholas, M.G. and Crispin, R.M. (1981) *Report AERE-R 10331*, AERE, Harwell.
8. Suganuma, K. and Okamoto, T. (1987) *Stud. Phys. Theor. Chem.*, **48**, 71.

9. Turwitt, M., Elssner, G. and Petzow, G. (1987) In *Ceramic Microstructures '86: Role of Interfaces* (eds J.A. Pask and A.G. Evans), p. 981, Plenum, New York.
10. Turwitt, M., Elssner, G. and Petzow, G. (1985) *J. Phys (Paris)*, **46**, C4–123.
11. Elssner, G., Suga, T. and Turwitt, M. (1985) *J. Phys. (Paris)*, **46**, C4–597.
12. Fischmeister, H.F., Elssner, G., Gibbesch, B. and Mader, W. (1988) *MRS Meet. on Advanced Materials, Tokyo, June*, paper F5.5.
13. Suga, T. and Elssner, G. (1985) *J. Phys. (Paris)*, **46**, C4–657.
14. Dundurs, J. (1987) *J. Compos. Mater.*, **1**, 310.
15. Liebowitz, H. and Eftis, J. (1971) *Eng. Fract. Mech.*, **3**, 267.
16. Suga, T. and Elssner, G. (1985) *Z. Werkstofftech.*, **16**, 75, 122.
17. Gross, B., Srawley, J.E. and Brown, W.F. (1965) *NASA Tech. Note D-3092*.

9

The non-destructive evaluation of ceramic materials and joints

H. Reiter

9.1 INTRODUCTION

Engineering ceramics with their strong but brittle properties are very sensitive to small defects and as such present new challenges in the field of non-destructive examination (NDE). Unlike metals where the critical flaw size can be several millimetres, in ceramic materials it is usually less than 100 μm. As ceramics are often joined to metals or other ceramics by brazing or diffusion bonding, inspection of the joint or interface poses additional problems, as does the substrate–coating interface in thermally sprayed coatings. Adequate and rapid means of detecting flaws in this size range are not yet readily available, and at present most candidate methods are in their development stage.

As ceramic materials are generally non-magnetic and non-conductive, the methods which have shown the most promise have in the main been developments or refinements of well-established processes based on acoustics or X-rays. Recently, however, techniques exploiting the properties of thermal waves have shown considerable potential for the NDE of ceramic materials.

Although individual methods based on these techniques are too numerous to discuss, the basic principles will be outlined and the limitations with respect to ceramics and interfaces will be described.

9.2 ULTRASONIC TESTING

9.2.1 Basic acoustics

(a) Sound waves

Sound, which is the oscillation of atoms or molecules, travels as waves in liquids, solids or gases. The velocity of these waves depends on the elastic properties and density of the media through which they travel. The waves can take one of four forms: longitudinal or compression waves, transverse or shear waves, surface or Rayleigh waves and plate or Lamb waves. The latter waveform generally occurs in metals only a few wavelengths thick and will not be discussed further.

Longitudinal waves consist of compression and rarefaction of the atoms and molecules which vibrate in the direction of travel, whilst transverse waves, analogous to waves in a rope, have particles vibrating normal to the direction of the wave (Fig. 9.1). The propagation of transverse waves requires a strong attractive force between vibrating particles, and hence these waves cannot travel in a gas or liquid. Their velocity is generally about half that of longitudinal waves. Surface waves can propagate at boundaries in solids where the displacement of the atoms is a combination of pressure and shear deformation.

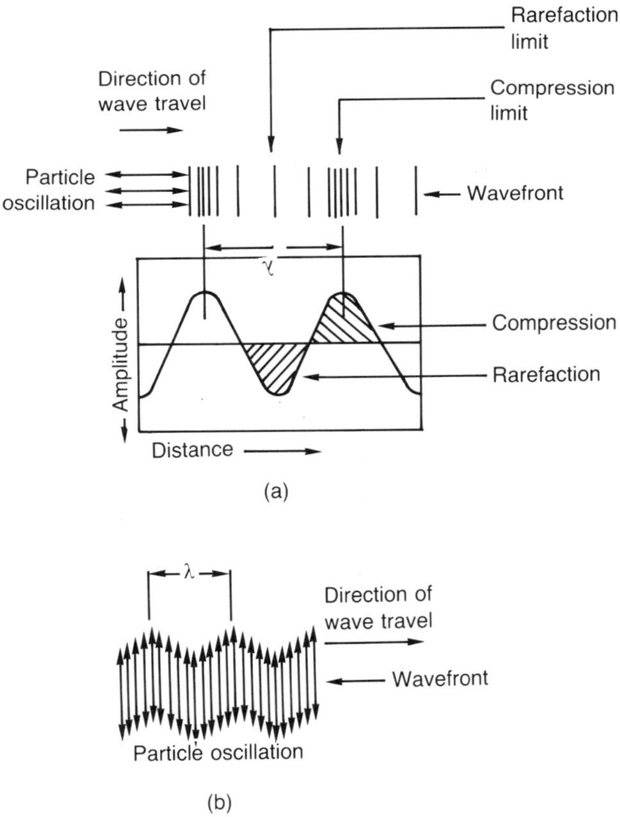

Fig. 9.1 Schematic representation of sound waves. (a) Longitudinal or compression waves (b) transverse or shear waves.

(b) Wave velocity

The velocity of a sound wave in a medium whose dimensions are much greater than a wavelength is dependent on the density of the medium and its elastic properties such that

$$V_c = \left(\frac{E}{\rho}\frac{1-\mu}{(1+\mu)(1-2\mu)}\right)^{1/2} \qquad (9.1)$$

$$V_s = \left(\frac{E}{2\rho(1+\mu)}\right)^{1/2} \qquad (9.2)$$

where V_c is the velocity of the longitudinal wave and V_s is that of the shear wave. E is Young's modulus, ρ is the density and μ is Poisson's ratio. The relationship between the frequency f, the wavelength λ and the velocity of propagation c for all types of ultrasonic waves except Lamb waves is given by the expression

$$c = f\lambda \qquad (9.3)$$

Consequently it should be remembered that the velocity of sound depends only on the medium through which it travels and that, when a wave of frequency f travels from one medium to another, a change in wavelength occurs.

(c) Attenuation

The intensity of sound at any point in a sound field decreases with distance from the source. This attenuation occurs for two main reasons: absorption of the sound energy by the medium and scattering of the sound beam.

Sound absorption decreases exponentially with distance such that

$$I = I_0 \exp(1 - 2A'x) \qquad (9.4)$$

where I_0 is the incident intensity, I is the intensity at a distance x from the source and A' is the absorption coefficient of the material.

Attenuation is caused primarily by the scattering of sound waves by individual grains making up the structure of the material or by any distributed porosity. Its relationship with frequency and grain diameter is complex and not well understood, but generally takes the form

$$a^* = a'f + b'D^3f^4 \qquad (9.5)$$

where a' and b' are constants and a^* is the attenuation. Thus the losses vary as the cube of the diameter D of the particle or pores, and as the fourth power of the frequency f. Therefore a coarse-grained material, which has a high attenuation coefficient, requires a low frequency, especially for large specimens, whereas much higher frequencies can be employed with fine-grained materials. The latter have the advantage of increasing the sensitivity of the ultrasonic imaging system. The spot diameter of a focused beam will at best be approximately equal to the wavelength of the ultrasonic wave, which in a solid at 10 MHz is about 0.5 mm, and hence it is desirable to use the highest frequency possible. Frequencies normally employed in NDE are in the range 5–25 MHz with higher values of up to 200 MHz being used in what is referred to as the acoustic microscope.

(d) Boundary and interface effects

Propagation of sound waves at a boundary between two media is similar to that of other waves in that reflection and refraction takes place. The only difference is that with acoustic waves partial mode conversion (from pressure to shear or vice versa) takes place when they strike the surface at an oblique angle.

The amount of energy transmitted or reflected at an interface depends on the acoustic impedance Z of the material where

$$Z = \rho V \tag{9.6}$$

and ρ is the density and V the velocity of the sound wave. The transmission property of an interface is determined by the difference or mismatch in the acoustic impedances of the two materials. Consequently, if the acoustic impedances are similar or equal there will be little or no reflection at the interface, whereas if the difference is large there will be little transmission. The amount of reflected and transmitted energy is related to the acoustic impedance of the materials by the relationships

$$a_R = \left(\frac{Z_1 - Z_2}{Z_1 + Z_2}\right)^2 \tag{9.7}$$

$$a_T = \frac{4Z_1Z_2}{(Z_1 + Z_2)^2} \tag{9.8}$$

where a_R and a_T are the reflection and transmission coefficients respectively and Z_1 and Z_2 are the acoustic impedances of the two materials. Thus air or gases which have a very low value of Z cause almost complete reflection when they form part of the interface.

From this brief outline of the behaviour of acoustic waves in elastic media it can be seen that changes in velocity, attenuation, transmission, reflection or frequency of the ultrasonic waves can be used to furnish information about the structure of a material.

9.2.2 Application of ultrasonic test methods

(a) Velocity and attenuation measurements

When the properties of a material are sensitive to overall structural features such as grain size, microporosity or microcracking etc., measurements of sound velocity or ultrasonic attenuation can be used for NDE of components.

Good correlation between sound velocity and bulk density has been reported for silicon carbide and silicon nitride [1,2], although significant changes in microstructure have only a relatively small effect on the actual ultrasonic velocity. However, variations in microstructure have a profound effect on the attenuation of high-frequency ultrasonic waves. The attenuation in large-grained silicon carbide was found to be much greater than in the fine-grained material with the same bulk density. It would thus appear that, because of the greater sensitivity to microstructure, attenuation measurements may be more useful in monitoring material strength.

(b) Reflection and transmission

Most conventional ultrasonic NDE is based on the transmission or back-reflection of ultrasonic waves to generate an acoustic image of a defect or second phase, with the contrast depending on the acoustic mismatch between the different media (Fig. 9.2). Since the speed of sound in ceramics is about twice that in metals, conventional ultrasonic equipment which generally uses frequencies of 5–25 MHz would give a wavelength of the order of 400–1000 μm. These wavelengths are too large for the resolution of flaws below 100 μm, which is the upper size for critical cracks in ceramics. To achieve resolution of this order, frequencies in the range 100–200 MHz are required. As the attenuation at these frequencies is high, only defects at or near the surface can generally be detected, although Hecht *et al.* [3] describe ultrasonic testing equipment working with a 25 MHz focused probe with which they examined hot-pressed Si_3N_4 (HPSN) and SiC plates 10 mm thick. They were able to detect areas of silicon enrichment, cracks and pores as deep as 4–6 mm.

Although inspection of welds in metals by pulse echo techniques is well understood, the application to diffusion bonds presents many difficulties. These have been reviewed by Derby *et al.* [4] and more recently by Kapranos and Priestener [5]. The problems with applying conventional ultrasonic pulse echo techniques mostly arise because the time of flight between any defect and the backwall echo is short and defects are normally

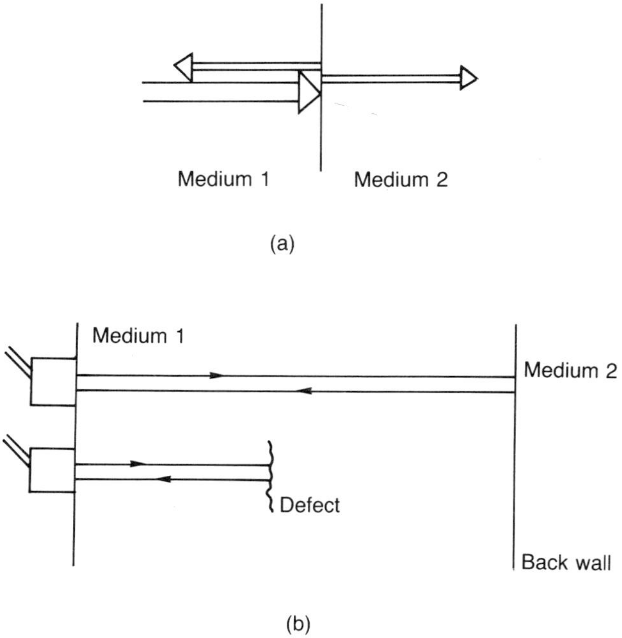

Fig. 9.2 Reflection and transmission of sound waves. (a) Transmission and reflection of sound waves at an interface (b) reflection of sound waves at a defect.

very small. A similar difficulty is encountered with thermally sprayed coatings which present a further problem in that their very porous and generally microcracked structure makes them highly attenuating to ultrasound, thereby limiting the thickness of coatings which can be examined to below 1 mm. However, both reflection and transmission techniques have been successfully used [6–8] to detect areas of debond at the metal–coating interface.

9.2.3 Ultrasonic spectroscopy

The attenuation due to multiple scattering in a material containing cracks, microporosity and other inhomogeneities increases exponentially with frequency and therefore limits the frequency which can usefully be employed. At lower frequencies and with broadband pulses the shape of the reflection echoes will be significantly changed from the incident beam, making amplitude and time delay measurements less meaningful. Under such circumstances, ultrasonic spectroscopy, in which the ultrasonic pulse is resolved into its Fourier frequency components, appears to furnish a method capable of monitoring structures and defects in materials. Vary and Kautz [9] have shown in a theoretical analysis that microstructural features could be related to a transfer function which was defined in terms of the frequency dependence of the attenuation coefficient. Lamidieu *et al.* [10], using a 1 MHz signal, demonstrated that the transfer function of the frequency spectra of the reflected pulse could be used to follow structural changes and the formation of cracks in thermal barrier coatings subjected to thermal fatigue. The method has also shown promise in the examination of diffusion-bonded titanium [5] where the 'spectral signature' of the reflected pulses was used to detect the difference between the frequency spectra for structural scattering and defect reflection at the diffusion-bond interface. The method should also be applicable to diffusion-bonded ceramics. However, the analysis and interpretation would be more difficult for brazed joints because of the additional spectral reflectances of the braze material. Steffens and Crostack [7] have reported a somewhat similar method for examining thermally sprayed coatings where they were able to show a shift in the peaks of the frequency spectrum when comparing the spectra from flawed and unflawed areas of interface.

9.2.4 The acoustic microscope

The need for high frequencies and short wavelengths to detect defects below 100 μm has attracted methods based on the recently developed acoustic microscope in which the images are formed by the interaction of sound waves with the sample [11]. The image contrast results from the differences in the elastic properties of the various constituents in the sample and can thus yield quite different information from the conventional optical microscope.

(a) Scanning acoustic microscopy

In the acoustic microscope (Fig. 9.3) the ultrasonic beam is focused on the surface by refraction through the coupling media and either the lens or sample can be rastered to build up an image from the reflection of longitudinal waves back into the coupling media, similar to the method used in conventional ultrasonic methods. Contrast is brought about through variations in the reflection coefficients across the sample, or by mode conversion of the incident acoustic beam to waves which propagate along the surface giving rise to interference patterns. The latter method is particularly sensitive to surface cracks [11].

This method is receiving increasing attention in the NDE field and shows great promise. An account of the application of a scanning acoustic microscope to alumina and silicon carbide samples is given by Nongaillard *et al.* [12] who were able to report the detection of defects to a depth of 5 mm with a spatial resolution in the region of 50–100 μm. Further examples are reported by Goebbels [13] for detection of porosity in

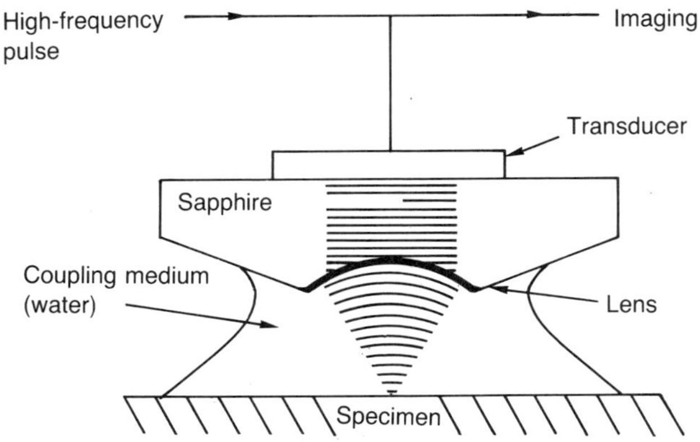

Fig. 9.3 Principle of reflection scanning acoustic microscope (SAM).

reaction bonded silicon nitride (RBSN) material to a depth of 12 μm using a frequency of 850 MHz. Smith [14], in an account of the possible applications of the acoustic microscope, describes the detection of discontinuities in a diffusion bond between a ceramic ring and a metal plate.

9.3 RADIOGRAPHY

9.3.1 General principles

X-rays are electromagnetic waves with a wavelength of about 1 Å (10^{-10} m) or less. At these wavelengths reflection and refraction only occur through very small angles, and for all practical purposes the rays can be considered to travel in straight lines from the source until they are absorbed or scattered by the media through which they travel.

A radiograph is produced by the shadow of a partially absorbing body upon a photographic film. The method is based on the differential absorption of the electromagnetic waves within the component caused by differences in density, variations in thickness or composition etc. Unabsorbed radiation passing through the part can then be recorded on photographic film, on a sensitized screen or by digital image processing.

(a) Absorption coefficient

The intensity of X-rays varies with thickness in the familiar form

$$I^* = I_0^* \exp(-\alpha^{**}x) \qquad (9.9)$$

where I_0^* is the intensity of the incident beam, I^* is the intensity of the transmitted beam and α^{**} is the linear absorption coefficient for the material. The numerical value of α^{**} depends on the wavelength of the incident beam and the material of the absorber.

(b) Absorption and scattering processes

When X-rays pass through matter three processes take place: the photoelectric effect, pair production and Compton scattering. The two former effects are effectively the absorption of a photon by atoms of the absorber accompanied by the production of one or two electrons, and Compton scattering is essentially the deflection of the photon by an atomic electron without much loss in energy. The relative importance of these different processes depends upon the wavelength of the radiation and the atomic number of the absorber. At low energies the photoelectric effect is dominant, whereas at intermediate energies Compton scattering is important with pair production occurring at above 1 MeV. The mass absorption coefficient due to the photoelectric effect is roughly proportional to the fourth power of the atomic number (hence the use of lead as a shielding material), pair production is roughly proportional to the atomic number and the Compton effect is little affected by the atomic number.

As X-rays travel in straight lines, a radiograph is produced in the same way as a shadow cast on a screen by an opaque body placed in the path of a beam of light. This shadow is shown in Fig. 9.4. It consists of a full

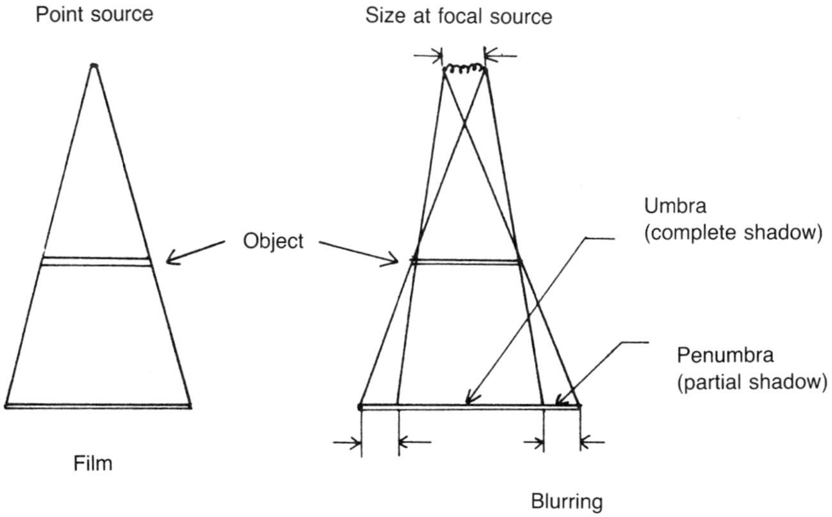

Fig. 9.4 Schematic representation of the effect of size of focal point on X-ray image. (a) Point source (b) finite length of source.

shadow (umbra) of the object surrounded by a region of partial shadow (penumbra). To obtain a sharper definition of the object the penumbra region must be reduced. This can be achieved by having the source as remote as possible, the source as small as possible or the object as near as possible to the film. Conventional X-ray sources have focal spots of diameter of the order of 2–5 mm, which makes the detection of defects of size less than 1 or 2 mm impractical. In order to achieve better resolution the focal point of the X-ray source must be reduced to as near a point source as possible and image enlargement achieved as indicated in Fig. 9.4.

9.3.2 Microfocal X-ray methods

Advances in X-ray equipment have enabled focal spot sizes as small as 1 μm with image magnification of 300 to be achieved [15], making the detection of small flaws possible. The probability of void or flaw detection will clearly also depend on the relative absorption coefficients of the various constituents making up the sample, whether they are voids or separate phases, in relation to the overall X-ray absorption of the matrix. Work reported by Klima *et al.* [16] showed that voids with diameters from 20 to 500 μm could be detected in Si_3N_4 in both green and fully sintered test bars with a sharp cut-off in detection probability below about 2% thickness sensitivity.

Unless the second phase or defect parallel to the direction of the beam is sufficiently thick to give a significant difference in absorption of the incident X-ray beam there will be insufficient contrast for defect detection. This makes the finding of planar defects very difficult and presents a particular problem with diffusion bonds, brazed joints and thin coatings. However, void detection in diffusion bonds has been reported by Goebbels [13] and Alreja [17].

9.4 THERMOGRAPHY AND THERMAL WAVE INTERFEROMETRY

9.4.1 Thermography

For many years thermographic techniques have been well-established methods for NDE of heat losses or hot spots, flaws and inhomogeneities in components etc. Information on such defects is obtained from observation of the surface temperature distribution during or after heating the component. The manner of heating can vary from a continuous heating source such as hot gases to a pulsed laser beam or flash bulb. Traditionally the methods relied on steady state temperature conditions being present in the component under inspection, but such conditions are usually difficult to achieve. However, transient changes in temperature, such as those that occur when a heat pulse diffuses through a material, can be used for detecting differences in thermal properties which can be correlated with defects and inhomogeneities within a component.

9.4.2 Pulsed video thermography

The method is based on the technique developed for measuring the thermal diffusivity of materials in which one surface of a thin disc is subjected to a thermal pulse from a flashlight and the rate of temperature rise is measured on the opposite face [18]. From the time–temperature curve thus obtained and the basic thermal diffusion equation, the thermal diffusivity of the material can be determined from the relationship

$$t^* = \frac{1.38L'^2}{\pi^2\delta^*} \quad (9.10)$$

where δ^* is the thermal diffusivity, L' is the thickness of the disc and t^* is the time to reach half the maximum temperature.

The rapid advances in infrared imagers and optical scanners has led to the development at the Harwell Laboratory [19] of a method known as pulsed video thermography, which is based on the above method of determining diffusivity. Measurements are made of the effect on the diffusion of a heat pulse through a body of differences in the diffusivities of the various phases, cracks, delaminations etc. which may be present. The change in the temperature transients at the surface is monitored by a thermal imager and recorded on a video cassette recorder for subsequent analysis.

Figure 9.5 shows a schematic diagram of the arrangement of the method. A thermal pulse strikes the surface of the material and diffuses through uniformly until it meets a thermal discontinuity (cracks etc.). As the heat can no longer diffuse freely through the discontinuity it will be dissipated more slowly, thereby generating different temperature transients compared

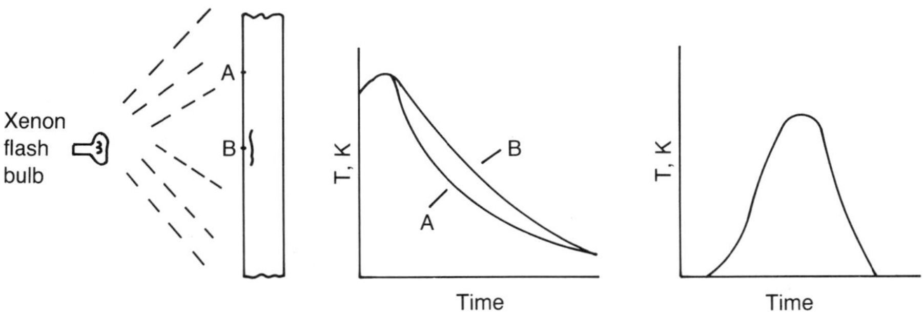

Fig. 9.5 Schematic representation of pulsed video thermography.

with the sound sections of the component. The high energy rate inputs of short duration required for materials with short thermal diffusivity half-lives were provided by xenon flash tubes.

The method, which was originally developed for the examination of carbon-fibre-reinforced plastic laminates and bonded joints, has been found to have great potential in other fields of NDE. The detection of subsurface defects in ceramic diffusion bonds and thermally sprayed coatings has been shown to be feasible [19]. The non-contact, single-sided and rapid nature of this method make it extremely attractive as a method for NDE and it will no doubt find wider application when the present very high cost of the imaging system comes down.

Another system which relies on the monitoring of thermal transients in which the heating source is a stream of hot gas has been described by Green [20,21]. The thermal decay transients are monitored immediately behind the heat source as indicated in Fig. 9.6. The presence of cracks or inclusions with different thermal properties will affect the rate of thermal decay and hence furnish information on the structure and quality of the

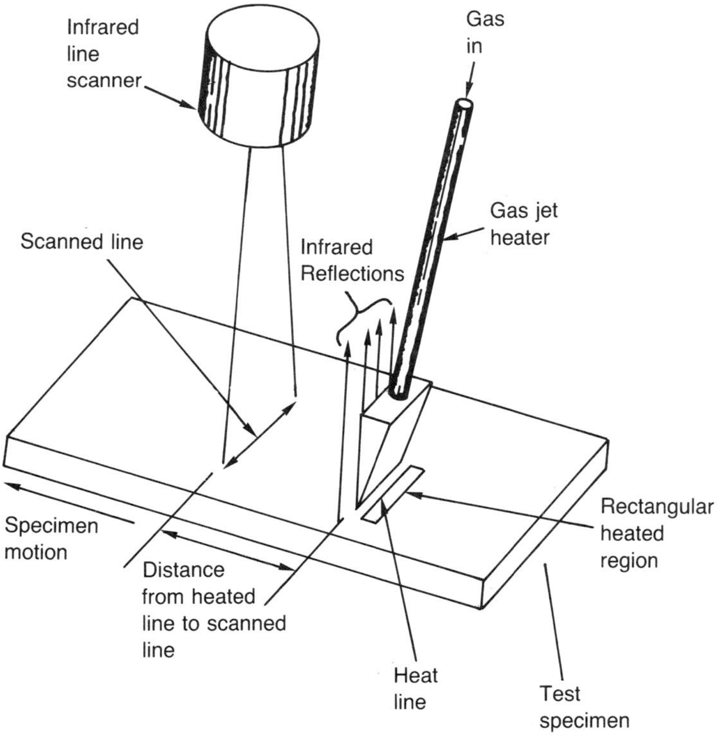

Fig. 9.6 Schematic representation of a pulsed thermal wave system.

material. The method has been used successfully to detect areas of debond in spray coatings and cracks and porosity in ceramics.

9.4.3 Thermal wave interferometry

When a periodic heat source is applied to the surface of a material, a heavily damped thermal wave is generated within the sample which can be represented by the equation

$$\Gamma = \Gamma_0 \exp\left(\frac{x}{l'}\right) \exp\left\{ \mathrm{j}\left(vt - \frac{x}{l'}\right)\right\} \tag{9.11}$$

where Γ_0 is the initial amplitude of the wave, ω is the angular frequency of the wave propagation in the direction x and l' is the thermal diffusion length which is defined as

$$l' = \left(\frac{k}{\rho C_{\mathrm{H}} \pi f}\right)^{1/2} \tag{9.12}$$

Here k is the thermal conductivity, ρ is the density and C_{H} is the specific heat.

If should be noted that, although heavily damped, a thermal wave is a propagating wave whose phase at successive points within the body lags the incident beam. Consequently information on the phase lag of the emitted signal can also be used to furnish information about the component in the same way as amplitude monitoring, with the additional advantage that it is insensitive to surface blemishes which could affect absorption of the incident energy [22]. The thermal diffusion length determines the depth of penetration of thermal energy into the material (Fig. 9.7). Thermal waves undergo reflection or scattering when they encounter regions of different thermal characteristics in the same way as light or ultrasonic waves do at regions of different refractive index or acoustic impedence.

The fraction R_{F} of a thermal wave that is reflected at an interface between two different media varies with the thermal mismatch and is given by

$$R_{\mathrm{F}} = \frac{1 - b^*}{1 + b^*} \tag{9.13}$$

where $b^* = p_2 c'_2 k_2 / p_1 c'_1 k_1$ and p, c' and k refer to the density, specific heat and thermal conductivity of the materials respectively.

A thermal wave generated in a material will be partially reflected at an interface with another material having different thermal properties, and

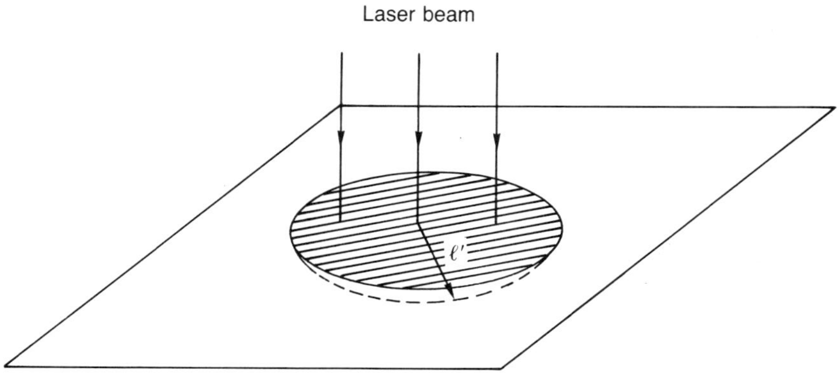

Fig. 9.7 The volume heated by a modulated light source, where μ is the diffusion length.

will return to the surface to produce interference effects. Bennett and Patty [23] have analysed the thermal wave interference phenomena produced by a coating and have shown that the amplitude and phase angle vary with coating thickness. Because of the heavily damped nature of the thermal wave this variation is confined to coatings whose thickness is comparable with the thermal diffusion length. This condition can usually be satisfied by the appropriate selection of modulation frequency. In this way it is possible to assess coating thickness [24].

In the same way thermal wave reflection will occur at adhesion defects between a coating and its substrate or at a subsurface crack in a component where there is effectively a gas interface. In such areas the amplitude and phase angle of the interference will alter since they depend on the reflection coefficient.

The potential value of thermal wave interferometry for NDE of components or the imaging of subsurface features has been discussed by a number of workers [25–27]. A typical laser imaging system developed at Bath University (Fig. 9.8) was successfully used to detect debonds in nickel chrome carbide and thermal barrier zirconia coatings up to 2 mm thick. This upper limit in thickness was due to the limitations of the equipment where the lowest frequency achievable was 0.5 Hz although at this frequency scanning rates become very slow. The system has also been successfully used to detect vertical cracks [28] at or near the surface as shown in Fig. 9.9. The presence of a crack hinders the lateral diffusion of heat, thereby increasing locally the amplitude of the emitted thermal waves which are then monitored by the infrared detector. This is an extremely sensitive system of crack detection with a resolution of three times the diffusion length, but in common with other scanning systems it suffers because of the time required to interrogate the surface.

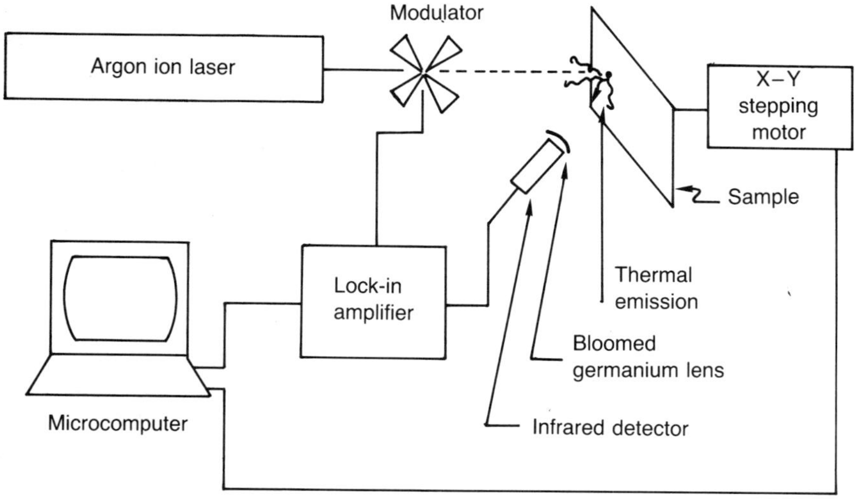

Fig. 9.8 Schematic representation of a thermal imaging system.

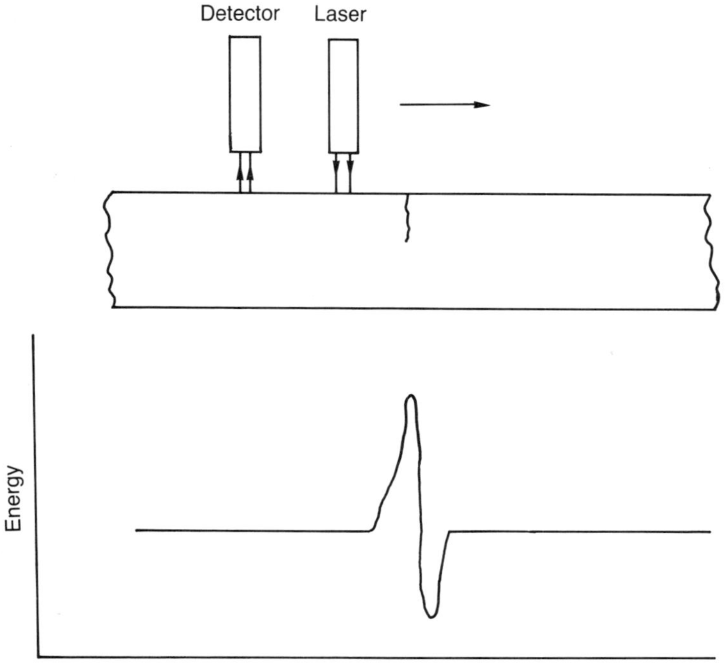

Fig. 9.9 Detection of vertical cracks by laser scanning.

One of the disadvantages of the early equipment was not only the size of the argon ion laser but the requirement of an adequate supply of cooling water as well as three-phase power. The recent availability of 0.5 W semiconductor lasers of negligible size with integrated power supply and control systems has made the construction of portable equipment possible. The development of optical fibres able to transmit infrared radiation has also opened up the possibility of getting away from the line-of-sight limitations of most current inspection systems [29].

9.4.4 Photoacoustic microscopy

The principle of the photoacoustic microscope is the same as that described in the previous section on thermal wave interferometry. The difference lies in the method of detection or imaging of the emitted thermal waves [30,31]. A modulated laser beam is again used to scan the sample, but through a window of a gas-filled acoustic cell surrounding the component. The temperature variations caused by the emitted thermal waves set up pressure waves inside the cell which are then monitored by a microphone and displayed in any of the usual ways. The method has been successfully applied to SiC blades where surface cracks and subsurface defects to a depth of 1 mm were readily detected [27,32,33] and defect resolutions of 30 μm or better have been claimed. The method suffers from the inconvenience that acoustic cells have to be built to contain the components or areas to be examined.

9.5 CONCLUDING REMARKS

Techniques for the non-destructive evaluation of ceramic–metal joints have yet to be widely used. This reflects basic difficulties in detecting flaws at interfaces between dissimilar materials where property changes are marked. However, development work aimed at applying techniques found to be suitable for flaw detection in ceramics is showing some promise and effective work has been done with coating–substrate interfaces. The next few years should see significant advances, but it is probable that these will be a refinement of existing ultrasound, X-ray and thermal methods rather than the introduction of new interogative techniques. These advances will reflect both increasing laboratory sophistication and the harsh imperatives of user demands as ceramics enter more demanding application areas.

REFERENCES

1. Kupperman, D.S. (1981) *Ultrasonic Characterization of Structural Ceramics; Analytical ultrasonics for structural materials*. (ed. A. Varey) Noyes, 99.
2. Klima, S.J. and Baaklini, G.Y. (1987) *Ultrasonic Characterization of Structural*

Ceramics; Analytical ultrasonics for structural materials. (ed. A. Varey) Noyes, 99.

3. Hecht, A., Neumann, E. and Rose. P. (1986) Computer aided ultrasonic testing of ceramic materials. *Proc. 2nd. Int. Conf. on Ceramic Materials for Engines*. Lubeck, DKG Bad Honef, FRG.
4. Derby, B., Briggs, G.A.D. and Wallach, E.R. (1983) Non-destructive testing and acoustic microscopy of diffusion bonds. *J. Mat. Sci.* **18**, 2345.
5. Kapranas, P. and Priestner, R. (1987) NDE of diffusion bonds. *Metals and materials* **3**, 194.
6. Frank, W. and de Gee, A.W.J. (1978) A non-destructive method for the adhesive bond strength of thermally sprayed non-fused coatings. *Proc. Conf. Advanced Surface Coating Technology*. Inst. of Welding, London, 99.
7. Steffens, H.D. and Crostack, H.A. (1980) NDT of thermally sprayed coatings. *Proc. 9th. Int. Conf. on Thermal Spraying*. The Hague, May, 120.
8. Cox, R.L., Almond, D.P. and Reiter, H. (1980) *NDT Int.* **13**, 291.
9. Vary, A. and Kautz, H.E. (1987) Transfer function concept for ultrasonic characterization of material microstructure. In *Material Analysis by Ultrasonics* (ed. A. Vary) Noyes, p. 249.
10. Briggs, G.A.D. (1985) *An Introduction to Scanning Acoustic Microscopy*. Oxford University Press, Oxford.
12. Nongaillard, B., Logette, P. and Rouvaen, J.M. (1984) New results in the application of ultrasonic microscopy to non-destructive evaluation. *Symp. on ultrasonics. I.E.E.E.*, Dallas, Nov.
13. Goebels, K. (1986) Non-destructive evaluation of ceramic components for engines. *Proc, 2nd. Int Conf. on Ceramic Materials for Engines*. DKG Bad Honef, FRG.
14. Smith, G.C. (1986) The scanning acoustic microscope – a new tool for the materials scientist. *Mat. Science and Technology*, **2**, 881.
15. Goebels, K. and Reinhold, A. (1986) Advances in microfocal X-ray equipment and state of the art in high resolution defect detection in ceramics. *Proc. 2nd. Int. Conf. on Ceramic Materials for Engines*. DKG Bad Honef, FRG.
16. Klima, S.J., Baaklini, G.Y. and Roth, D.S. (1986) Factors affecting reliability of non-destructive detection of flaws in structural materials. *Proc. 2nd. Int. Conf. on Ceramic Materials for Engines*, DKG, Bad Honef, FRG.
17. Alreja, V. (1984) Non-destructive evaluation of diffision bonded Ti using real-time radiography. Munich, 933.
18. Parker, J.W., Jenkins, R.J., Butler, C.P and Abbot, G.L. (1981) Flash method of determining thermal diffusivity, heal capacity and thermal conductivity, *J. App. Phys.*, **32**, 1679.
19. Reynolds, W.N. and Wells, G.M. (1984) Video–compatible thermography. *Brit. J. of NDT*, **26**, 40.
20. Green, D.R. (1971) Thermal and infrared NDT of composites and ceramics. *Mat. Evaluation*, **12**, 242.
21. Green, D.R., Schmeller, M.D. and Sulit, R.A. (1982) Thermal NDE method for thermal spray coatings. *Proc. Oceans Conf.*, Washington DC.
22. Busse, G. (1982) *App. Optics*, **21**, 107.
23. Bennett, C.A. and Patty, R.R. (1982) Thermal wave interferometry: a potential application of the photo-acoustic effect. *Proc. Oceans Conf.*, Washington DC. p. 49.
24. Almond, D.P., Patel, P.M. and Reiter, H. (1985) An evaluation of the suitability of thermal wave interferometry for testing plasma sprayed coatings. *NDT Int.*, **18**, 17.

25. Jaarin, J. and Luukkala, M. (1984) The use of thermal wave for NDE of sprayed coatings. *3rd. European Conf. on NDT*, Florence, Italy.
26. Cielo, P. (1983) Analysis of pulsed thermal wave inspection. *Proc. 14th Symp. on NDE*, San Antonio, 418.
27. Hoffman, B., Grise, H. and Arnold, W. (1986) NDE of ceramics by photo-acoustic and high frequency ultrasonic methods. *Proc. 2nd. Int. Conf. on Ceramic Materials for Engines*, DKG Bad Honef, FRG.
28. Bath University (UK), private communication.
29. Bath University (UK), private communication.
30. Rosencwaig, A. (1982) Thermal wave Microscopy. *Solid State Tech.*, **25**, 91 (1982).
31. Thomas, R.L. (1980) *J. Appl. Phys.*, **51**, 1152.
32. Goebels, K., Reiter, H., Hirsekorn, S. and Arnold, W. (1984) *Sci. of Ceramics*, **12**, 483.
33. Khandelwal, P.K., Kinnick, R.R. and Heitman, P.W. (1985) Cer. Bulletin, **8**, 1112.

10

Joining silicon nitride to metals and to itself

K. Suganuma

10.1 INTRODUCTION

Among emerging engineering ceramics, silicon nitride is a candidate material with an outstanding potential for high-temperature structural applications. It has a high failure strength, hardness and elastic modulus, good wear, oxidation and thermal shock characteristics and is a light-weight material. Silicon nitride does not melt but decomposes at about 1870 °C in nitrogen. It has two crystallographic forms, the α and β phases, which are both hexagonal but differ in lattice spacing with densities of 3.18 $g\,cm^{-3}$ and 3.19 $g\,cm^{-3}$ respectively.

Commercially manufactured silicon nitrides are mostly β phase and can be classified into three groups: oxide-bonded pressureless sintered (PLS), silicon reaction bonded (RB) and pure silicon nitride with little or no sintering additive such as hot-pressed (HP), hot-isostatically-pressed (HIP) and chemical vapour deposited (CVD) material. PLS silicon nitrides contain oxides such as alumina, calcia and yttria up to 10–15 wt% as sintering additives. (PLS silicon nitride could be extended to sialon ceramics which consists of Si–Al–O–N.) These ceramics have strengths of 600–1000 MPa which are maintained up to 1000–1200 °C. RB silicon nitrides contain free silicon, their relative density is about 60%–85% and they have a strength of about 200 MPa up to about 1500 °C. However, oxidation becomes severe beyond 1000 °C because of the formation of silica from free silicon. HP silicon nitrides contain a little oxide phase and keep their strengths of 800–1200 MPa up to 1200 °C. CVD silicon nitride, which is the purest and has no grain boundary phase, maintains a strength of 400 MPa even above 1500 °C.

Although conventional joining methods for ceramics have concentrated on oxide ceramics, especially for electrical components, a considerable number of joining techniques were developed and reported for silicon nitride between 1984 and 1986. This reflects its current importance. When attempting to establish a sound and tight joining process for ceramic–metal components, two basic factors must be considered for every application: interfacial strength and expansion mismatch.

Strong ceramic–metal interfaces can be achieved by both chemical bonding and lattice matching. The reaction conditions must be selected to optimize the interfacial strength for each combination of constituents. For this strength to be maintained it is necessary to relax the thermal expansion mismatch between the ceramic and metal. Ceramics have relatively small thermal expansion coefficients compared with most metals, and silicon nitride has a particularly small thermal expansion coefficient even compared with other engineering ceramics such as silicon carbide, alumina and zirconia. The thermal expansion coefficient of silicon nitride is about $3 \times 10^{-6}\,°C^{-1}$, while those of the steels, which are now used as structural components, are greater than $14 \times 10^{-6}\,°C^{-1}$. This difference in thermal expansion coefficient can cause great damage to a joint. For example, after silicon nitride and a steel are joined at 1000 °C, cooling to room temperature produces a 1% mismatch, which severely weakens the joint because of the large residual stresses generated. Therefore modification of joint designs is required. In overcoming this problem, one Japanese motor company succeeded in producing a silicon nitride bladed turbocharger witl a steel shaft (Fig. 10.1) in a collaborative development programme with ; Japanese ceramic manufacturer [1].

In this chapter we are concerned first with interfacial reactions anc microstructures between silicon nitride and metals, secondly with th thermal expansion mismatch effect, thirdly with other important factor such as surface roughness and stress corrosion, and finally with joinin; silicon nitride to itself with or without oxide interlayers.

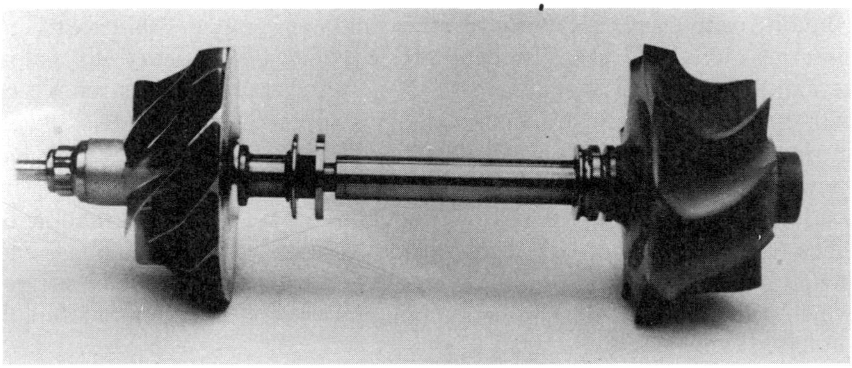

Fig. 10.1 Turbocharger rotor with a silicon nitride blade (right-hand side). The blade is joined to a steel shaft using laminate interlayers and active metal brazing [1].

10.2 INTERFACIAL REACTIONS AND MICROSTRUCTURES FORMED BETWEEN SILICON NITRIDE AND METALS

Interfacial microstructures between silicon nitride and metals can be divided into three groups, depending on their reaction bonding processes. The first group is the interfaces made by active metal brazing, one of the most popular joining processes for ceramic–metal systems including oxide ceramics. The second group is interfaces formed by a eutectic melting reaction between silicon nitride and metals. Silicon ceramics including silicon carbide react with certain transition metals to form eutectic liquids, and this method utilizes this reaction by optimizing the conditions. The last group are interfaces formed by solid state reactions under pressure.

The reaction process and products for each group can be partly predicted from thermodynamic data. For instance, let us suppose that the reaction between silicon nitride and a metallic element M_e produces a metal silicide and free nitrogen:

$$Si_3N_4 + M_e \rightarrow M_e \text{ silicide} + \text{free N} \tag{10.1}$$

Free energy data relevant to some of these reactions are shown as a function of temperature in Fig. 10.2 [2]. The resultant free nitrogen may

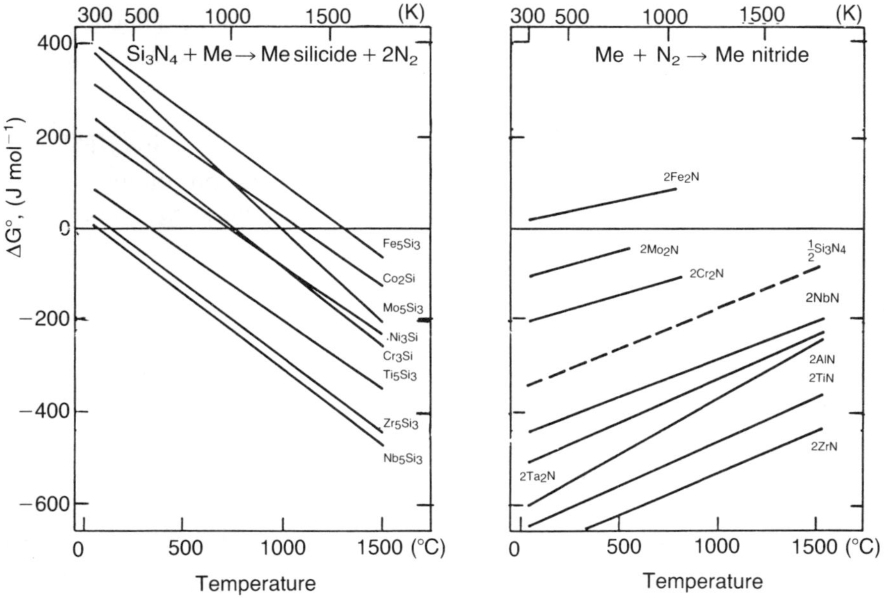

Fig. 10.2 Free energies of formation of metal silicides by reaction between silicon nitride and metals, and of metal nitrides [2], as a function of temperature.

react with the metal to form a metal nitride, remain in it as nitrogen gas or solute, or be released from the system as nitrogen gas. In the temperature range in which the free-energy change is negative, the reaction proceeds to the right-hand side of the equation.

If the formation of a metal nitride becomes the rate-determining step, the reaction

$$Si_3N_4 + M_e \rightarrow M_e \text{ nitride} + \text{free Si} \tag{10.2}$$

will dominate. The free-energy change of this reaction can be obtained from Ellingham diagrams, as shown in Fig. 10.2. The released free silicon may also react with the metal. For niobium, titanium, tantalum, aluminium and zirconium, the reactions can occur over the entire temperature range of interest when bonding because their nitrides are more stable than silicon nitride.

However, it must be noted that the minimum temperatures for both reactions of equations (10.1) and (10.2) are important, and cannot be deduced from the free-energy changes alone because at low temperatures the diffusion and reaction rates are sluggish. In addition, the formation of nitrides also depends on the partial pressure of nitrogen in the bonding environment.

Nevertheless, Fig. 10.2 provides useful suggestions for the selection of reactive elements that can promote bonding to silicon nitride. In the case of PLS and RB silicon nitride, bonding characteristics will be influenced by the presence of grain boundary phases, i.e. free silicon and oxides. Free silicon will react with many metals to form silicides, and oxides may promote wetting by strong oxide formers.

Several examples of successful joining methods are described below. Because interfacial microstructure and strength depend on joining conditions, especially joining temperature and time, in general the interfacial structures to be shown are those of the strongest joints.

10.2.1 Active metal brazing

The so-called active metal brazes contain or are themselves active elements. Examples include an Ag–Cu eutectic alloy with a few weight per cent addition of titanium or zirconium, a Cu–Ti eutectic alloy [3–7], pure aluminium and Al–Ti alloys [5,8–10]. The role of active metals in forming the interface is discussed in general elsewhere (Chapter 5), but this section concentrates on their use in bonding nitrides. The silicon nitride turbocharger rotor shown in Fig. 10.1 was produced by using one kind of Ag–Cu–Ti braze and several other examples are cited below.

Active metal brazing is performed without any metallization. Work-

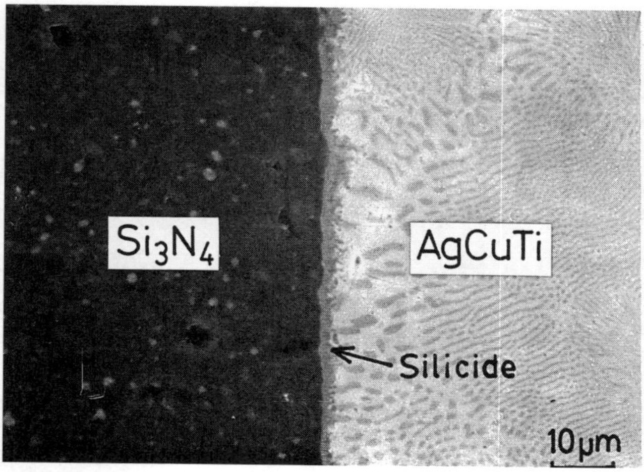

Fig. 10.3 Scanning electron micrograph of the interfacial microstructure of a silicon nitride–(Ag–Cu–Ti) braze reacted at 900 °C for 10 min in a vacuum. A thin silicide layer is observed but no titanium nitride because it is too thin.

pieces are jigged, usually applying a slight load, and held for a few minutes at the brazing temperature (800–950 °C for Ag–Cu–Ti brazes). The brazing atmosphere is a vacuum of less than 10^{-4} mbar. Under these conditions, bend strengths of up to 500 MPa can be achieved with silicon nitride. The interfacial microstructure of such a joint is shown in Fig. 10.3.

The formation of both nitride and silicide products is predicted by the free-energy diagrams shown in Fig. 10.2, and the main reaction products observed are titanium nitride adjacent to the silicon nitride and a complex outer eutectic layer. The titanium nitride layer, which consists of fine grains, is very thin (about 1000 Å [6]) while the complex eutectic layer consists of titanium silicide, mainly Ti_5Si_3, and the Ag–Cu–Ti eutectic.

Several modifications of the active metal brazing process are possible, e.g. Ag–Cu brazing combined with metallization of silicon nitride by vapour deposition of titanium or zirconium films, or brazing with a laminate of a titanium foil on an Ag–Cu foil. The microstructures obtained by these processes are similar to that shown in Fig. 10.3.

Aluminium is also an active element relative to silicon nitride, forming a stable nitride but not a silicide. Figure 10.4 shows the microstructure of silicon nitride and aluminium interfaces formed at 800 °C [11]. Two reaction layers are present: the one adjacent to the silicon nitride is about 500 Å thick and consists of fine grains of β'-sialon, while the other is an amorphous aluminia layer containing silicon about 4500 Å thick which ultimately forms mullite.

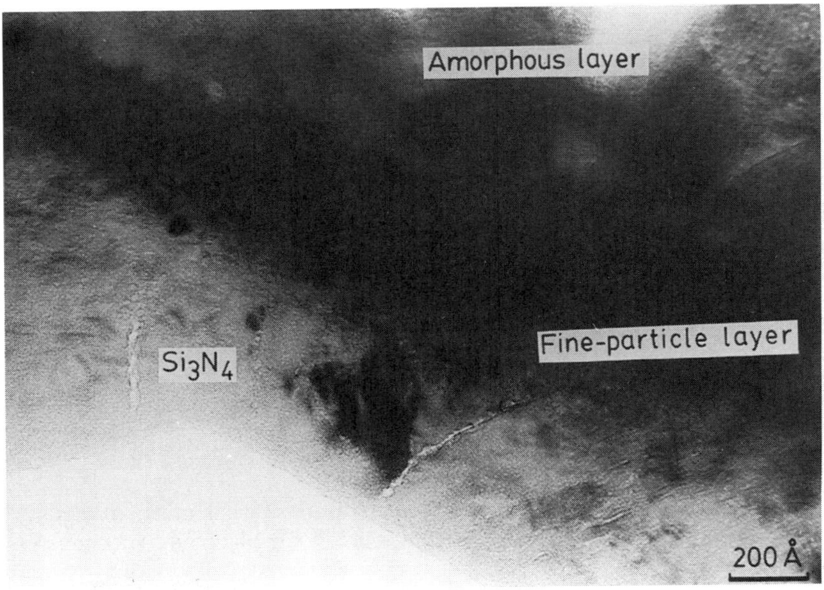

Fig. 10.4 Transmission electron micrograph of a silicon nitride–aluminium interface reacted at 800 °C for 15 min in a vacuum. The fine-particle layer consists of sialon grains smaller than 100 Å.

Above 1000 °C, 15R–AlN-type sialon was formed instead of β'-sialon [12]. The fact that sialon or amorphous Al–Si–O, both of which contain oxygen as one of the elements, was formed by the reaction indicates the importance of the partial pressure of oxygen during brazing. It is well known that silicon nitride is unwetted by aluminium at 800°C [13], but the presence of oxygen in the reaction should promote wetting. In fact, wetting was better in an argon atmosphere, which is slightly oxidizing, than in a vacuum [10]. Again, pre-oxidation of silicon nitride to form a silica layer on the surface promotes the wetting [14]. As with the Ag–Cu–Ti braze, the maximum bending strength obtained using aluminium as a braze was more than 500 MPa.

10.2.2 Solid state bonding

Refractory metals such as molybdenum, niobium, tantalum, zirconium and tungsten react with silicon nitride in the solid state. They also produce strong interfaces with silicon nitride when bonded at 1300–1500 °C. The main reaction products are silicides M_5Si_3 (M is the refractory metal atom) and refractory metal nitrides which are more stable than silicon nitride, as seen in Fig. 10.2. Other types of silicides were also recognized in some

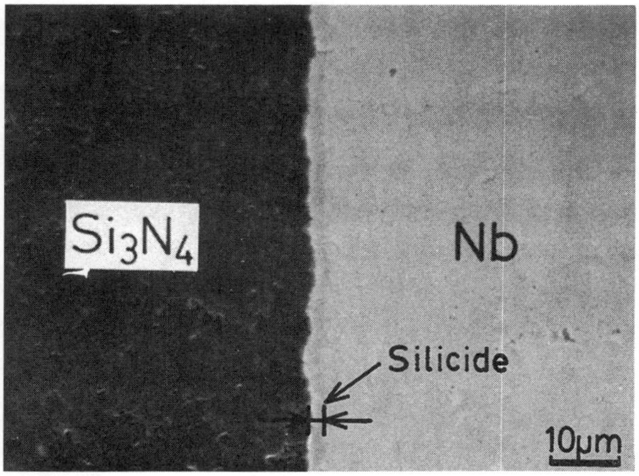

Fig. 10.5 Scanning electron micrograph of the interfacial microstructure of silicon nitride–niobium reacted at 1400 °C for 30 min under a pressure of 100 MPa. The niobium silicide layer is very thin.

cases [15, 16]. A typical interfacial microstructure is presented in Fig. 10.5. The nitride layer could not be recognized, but a silicide layer was observed.

Since thick silicide layers sometimes contain cracks, the strength of such a joint decreases. The maximum interfacial strength requires optimization of bonding temperature, time and pressure. In a solid state process, the pressure has a marked effect on the contact at an interface and this effect will be discussed elsewhere. Isostatically applied high pressure is most effective for completing the contact [17]. When conditions are optimized it is possible to obtain an interfacial strength above 300 MPa for silicon nitride–refractory metal systems.

10.2.3 Eutectic joining

Iron, nickel, cobalt, low alloy steels, austenitic and ferritic stainless steels, and nickel base superalloys react with silicon nitride to form eutectic liquids between silicon and the metal elements. The eutectic temperatures are about 1100 °C for nickel alloy–silicon nitride systems and about 1200 °C for iron alloy–silicon nitride systems [17–19] and may be decreased by the presence of oxygen [20]. The reaction products are mainly metal silicides such as Fe_5Si_3 which grow to a thickness of about 1000 Å, but a silicon diffusion layer is formed on the metal side. A typical interfacial microstructure is shown in Fig. 10.6. Metal nitrides of the iron system are not stable at the joining temperature when bonding is in a vacuum. The

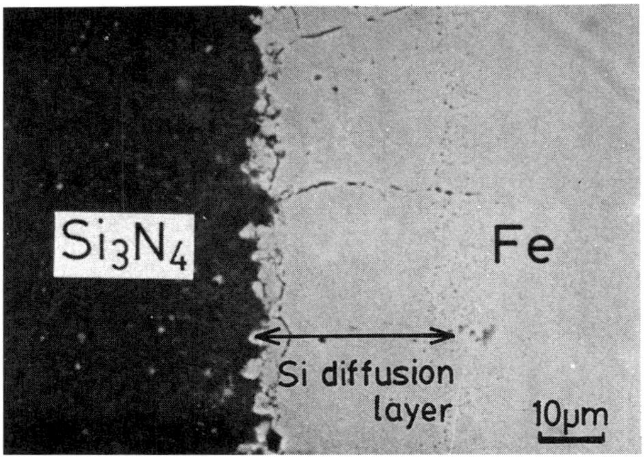

Fig. 10.6 Scanning electron micrograph of the interfacial microstructure of silicon nitride–iron reacted at 1200 °C for 30 min under a pressure of 100 MPa. Diffusion of silicon into iron is observed.

interfacial strengths can be more than 300 MPa for joints between iron, nickel and their alloys and silicon nitride.

10.2.4 Interfacial strength

The interfacial strengths obtained by optimum joining conditions for the various metals mentioned above are summarized in Table 10.1. From these data it is clear that these joining methods can produce a satisfactory interfacial strength if conditions are appropriately selected. Unfortunately, since the strength testing methods and the metal layer thickness were not identical for every case, an absolute comparison of strength was not available. However, these data provide some guidance.

Although aluminium lacks strength above 300 °C, it can produce strong joints at a low temperature with a simple joining process. Ag–Cu–Ti brazes have useful strengths at medium temperatures up to 600 °C, but do not have oxidation resistance. Eutectic joining materials can have useful strengths at slightly higher temperatures than active metals and also have oxidation resistance if they contain high levels of nickel or chromium. Thus Fig. 10.7 shows the oxidation resistance of silicon nitride–iron system joints at 800 °C in air. Joints produced using iron and type 304 stainless steel were not oxidation resistant, but those produced using nickel and type 410 stainless steel kept their strength levels of more than 100 MPa even after 100 h exposure at 800 °C in air. Solid state bonds with refractory metals can give useful strengths even at 900–1000°C but do not have

Table 10.1 Interfacial strength of silicon nitride–metal joints

Group	*Bonding metal*	*Joining condition*	*Metal thickness (mm)*	*Strength* (MPa)*
Active metal brazing	Ag–33.5Cu–1.5Ti	950 °C, 0.001 MPa, 60 min	15	650/800 [5]
	Cu–4Ti	1100 °C, 0.01 MPa, 60 min	–	300/–
	Al			370/470
	AA6061	850 °C, 3.3 MPa, 15 min	<10	470/470
	AA2017			400/470
	Al–(Al–10Si) clad	610 °C, 9.8 MPa, 30 min	600	300/500 [9]
Solid state bonding	Mo	1500 °C; 10–100 MPa, 30 min	100	420/500–600
	Nb	1400 °C; 10–100 MPa, 30 min	100	290/500–600
	Ta	1400 °C; 10–100 MPa, 30 min	100	380/500–600
Eutectic bonding	Fe	1200 °C, 10 MPa, 30 min	200	330/500–600
	Ni	1200 °C, 10 MPa, 30 min	200	250/500–600
	Steel type 410	1200 °C, 10 MPa, 30 min	200	320/500–600
	Steel type 304	1200 °C, 10 MPa, 30 min	200	220/500–600

All data were evaluated by bend tests.
*Strength (*A*/*B*): *A* is the joint strength and *B* is the strength of a silicon nitride or sialon.

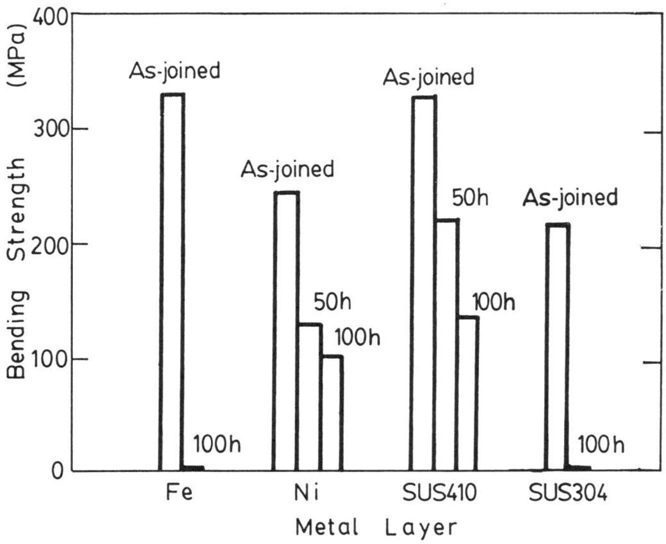

Fig. 10.7 Oxidation resistance of silicon nitride–iron joints. All the joints were joined at 1200 °C for 30 min under a pressure of 10 MPa in a vacuum and subjected to 800 °C exposure in air. Four-point bend strengths were measured at room temperature.

oxidation resistance. Thus each joining method has its own advantages. An appropriate method for the application must be selected.

10.3 RESIDUAL (THERMAL) STRESSES IN SILICON NITRIDE–METAL JOINTS

10.3.1 Outline of residual stress

The distribution of residual stress in a ceramic–metal joint is not uniform even along the interface. The residual stresses are greater closer to the interface or the free surface. The most harmful effect is caused by the tensile component of the residual stress at the interface or in the ceramic, and the maximum tensile stresses occur at or near the interface and the free surface, as shown in Fig. 10.8 [21]. The arrows indicate both the position and the direction of the maximum tensile stress. Since this tensile stress acts almost perpendicularly to the interface, the apparent interfacial strength measured by tensile or bend tests is considerably reduced. If the interfacial bond strength is large enough, the fracture will take place in silicon nitride, initiating at the point where the maximum tensile stress occurs in Fig. 10.8. Figure 10.9 shows a failed silicon nitride–iron joint; such 'concave/convex' fracture modes are observed in joints with large thermal expansion mismatches.

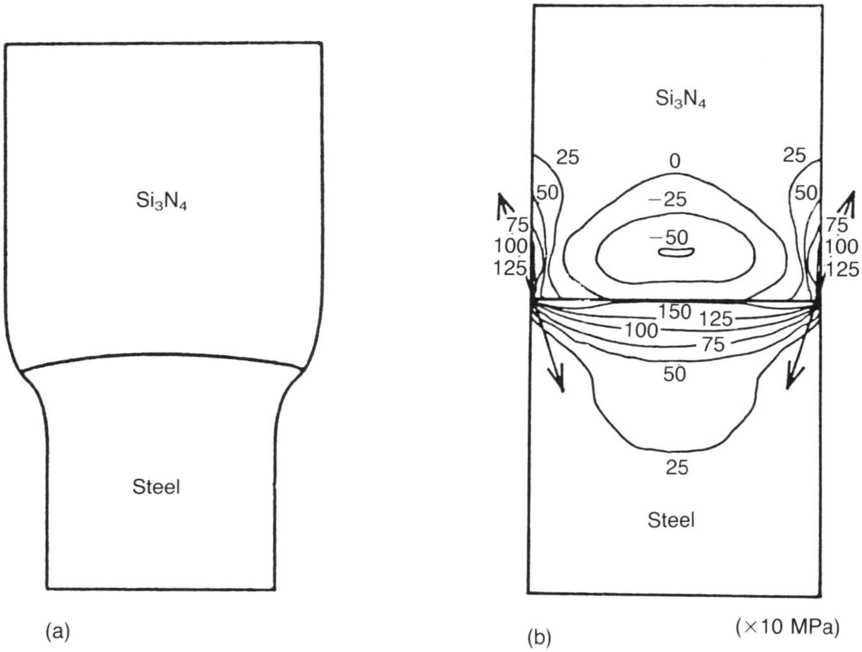

Fig. 10.8 (a) Deformed silicon nitride–steel joint and (b) distribution of maximum principal stress calculated by a finite-element method. The joint is assumed to be cooled from 1000 °C to room temperature and to deform elastically. The strain is enlarged by a factor of 25. The arrows indicate the position and direction of the maximum tensile stress.

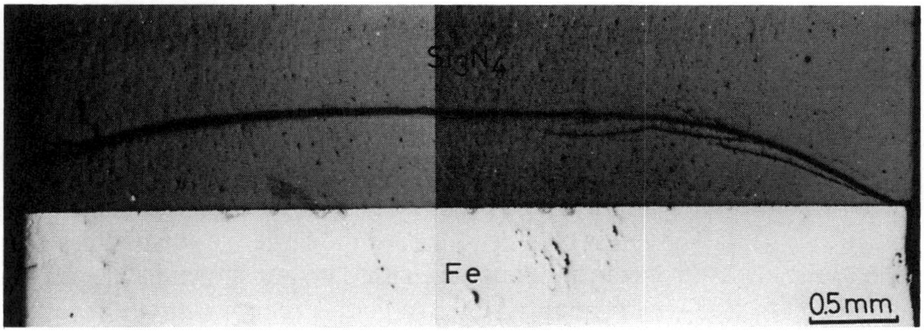

Fig. 10.9 Typical fracture of joints with large thermal expansion mismatch. In this case silicon nitride was joined to iron at 1200 °C for 30 min under a pressure of 100 MPa.

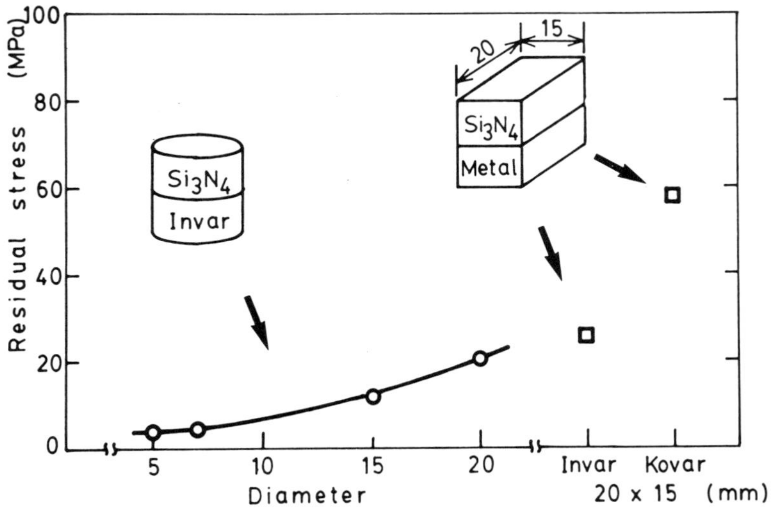

Fig. 10.10 Dependence of the tensile residual stress in silicon nitride near the interface on the diameter, shape and type of metal, as measured by a strain gauge method.

The magnitude of the residual stress depends on the shape and size of the interface [22]. Thus, Fig. 10.10 shows the diameter dependence of the surface residual stresses of cylindrical and rectangular face joints of silicon nitride and Invar alloy. The larger joints experience the larger residual stresses and stress concentrations at the corner of the rectangular joint are high.

A joint with large thermal expansion mismatch is generally weakened. However, sometimes one specimen is strong while another is weak, even if they are the same type. This effect is due to differing distributions of internal flaws induced by residual stresses during joining. To study this in more detail, the strengths of silicon nitride–Invar and silicon nitride–Kovar joints which differed in the magnitude of residual stresses were examined statistically [23]. The latter type of joint had the larger residual stresses as seen in Fig. 10.10, and the resultant strength data are shown in Fig. 10.11. The silicon nitride–Invar joint exhibited a good strength with little scatter, but the silicon nitride–Kovar joint had a complex strength dependence. About half the samples had a high strength with little scatter, while the others were weak with very variable scatter. The latter joints always had interfacial flaws which were formed during cooling from the joining temperature. Thus large residual stresses sometimes induce flaws in a joint, which not only weaken it but also can produce considerable scatter of the data. In other words, it is very important to examine the degree of scatter

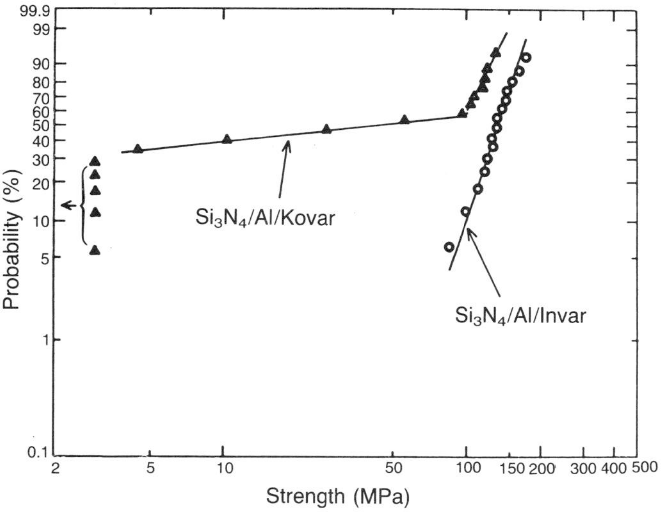

Fig. 10.11 Weibull plots of bend strengths of silicon nitride–Invar and silicon nitride–Kovar joints brazed with aluminium at 800 °C for 7 min in an argon flow (0.15 MPa).

in the strength of silicon nitride–metal joints because the residual stress can be large in this system.

10.3.2 Compensating for thermal expansion mismatch effects

To compensate for thermal expansion mismatch effects, it is necessary to reduce the tensile part of the residual stress in a ceramic or at an interface. This is because, although ceramics can withstand compressive stress, they are weak in tension. In contrast, the tensile stress in a metal can be accommodated by elastic and plastic deformation. A wide variety of methods using interlayers have been developed for ceramic–metal systems to promote mismatch accommodation, e.g. soft metal interlayers [24–27], composite interlayers (particulate and fibrous) [28,29], laminate interlayers (soft metal with a low-expansion hard metal) [30] and microcracked interlayers [10,31]. Details of these are mentioned elsewhere in this volume, and examples relevant to silicon nitride–metal joints are summarized in Table 10.2.

The soft metal interlayers such as aluminium and copper reduce the residual stresses by their elastic, plastic and creep deformation. However, such a single layer does not remove the residual stress efficiently for silicon nitride–metal systems. In addition, it would not resist sudden temperature changes or severe heat cycles because even for soft metals plastic and creep

Table 10.2 Interlayers used for silicon nitride–metal joints

Group	*Base metal*	*Interlayer*	*Joining condition*	*Strength* (MPa)*	*Reference*
Soft metal	WC–Co, Kovar, steel	Al-clad	610 °C, 9.8 MPa, 30 min	250–300 (b)	9
		Al-clad/Kovar	610 °C, 9.8 MPa, 30 min	200 (b)	9
		Steel type 304/WC	1100 °C, 5 MPa, 60 min	150 (b)	32
Soft metal–hard low-expansion metal laminate	Type 405	Fe–W	1200 °C, 100 MPa, 30 min	>60 (t)	17
	Mo	Mo–W	1500 °C, 100 MPa, 30 min	370 (b)	33
	Nimonic 80A		950 °C, 54 MPa, 30 min	100 (t)	34
Cracked layer	Type 405	Al–Invar	800 °C, 0.15 MPa, 7 min	150 (b)	10

*b, bend strength; t, tensile strength.

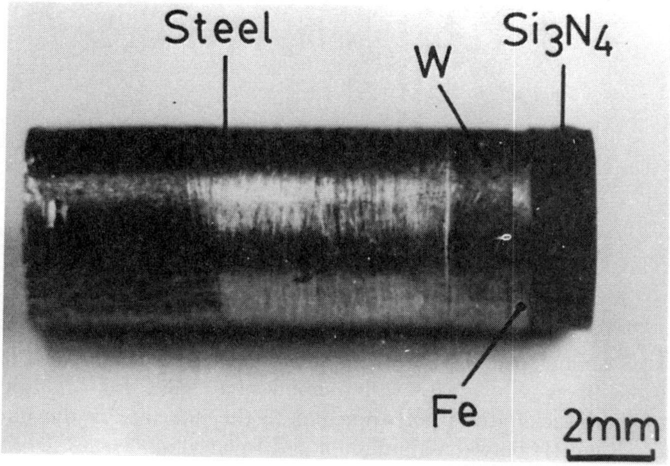

Fig. 10.12 Silicon nitride–type 405 steel joint with an Fe–W laminate interlayer. Silicon nitride and Fe–W were joined at 1200 °C for 30 min under a pressure of 100 MPa by HIP and were then brazed to the steel with a nickel base braze at 850 °C for 5 min in a vacuum.

deformation cannot respond to sudden temperature changes. However, a laminate interlayer can improve resistance to sudden temperature changes. Since the soft metal is restricted by silicon nitride and a low-expansion hard metal on its two sides it cannot deform readily following the shrinkage of silicon nitride. In addition the harmful effect of the large expansion or shrinkage of a base metal is blocked by the hard metal layer. Figure 10.12 shows an example of a silicon nitride–steel joint with a laminate interlayer of Fe–W. This joint had a strength of more than 60 MPa in tension even after thermal cycling between room temperature and 500 °C. The turbocharger rotor shown in Fig.10.1 was also produced with a laminate interlayer. It has a (soft metal)–(low-expansion hard metal)–(soft metal) interlayer structure with a total interlayer thickness of 4 mm. However, the laminate interlayer method still has problems. How to bond a large area is one of the more serious of these because the residual stress increases with bond area (Fig. 10.9) and ultimately causes fracture of the silicon nitride.

Composite interlayers such as silicon nitride–metal mixtures are another effective method of accommodating thermal mismatches, but a problem is the strength and reliability of the interlayer itself. Microcracked interlayers can produce a strong joint, and the silicon nitride–steel joint with an Invar interlayer bonded by aluminium brazing shown in Figs 10.10 and 10.11 has this type of interlayer structure. Figure 10.13 shows the interfacial microstructure. The network of fine cracks present perpendicular to the interface in the reaction layer between the aluminium and the

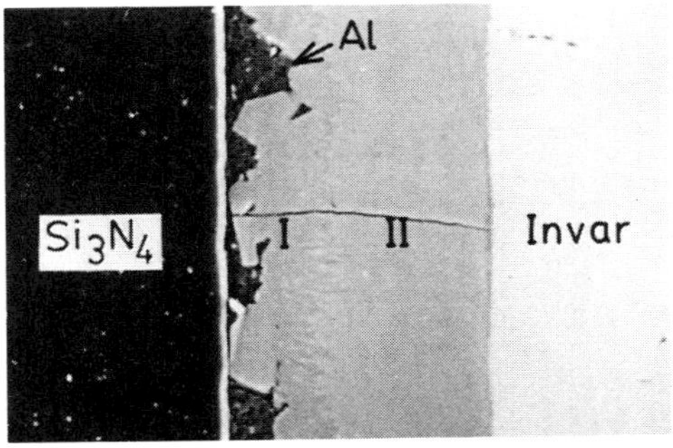

Fig. 10.13 Scanning electron micrograph of the interfacial microstructure of a silicon nitride–Invar joint brazed with aluminium at 800 °C for 7 min in argon. A network of fine cracks grows perpendicular to the interface in the intermetallic compounds (I and II) between aluminium and Invar alloy.

Invar alloy releases the residual stresses. In this type of interlayer it is necessary that the cracked layer is sandwiched between ductile layers to prevent the growth of cracks, particularly into the silicon nitride. In this case the residual aluminium layer plays this part.

10.4 OTHER IMPORTANT FEATURES

The surface roughness and surface preparation technique can influence joint properties, particularly strength. Thus a rough surface will prevent complete contact during diffusion bonding using low pressures and the surface layer of the ceramic near an interface may have deep scratches which can cause severe residual stressing. However, a rough bond face may have an anchoring effect, promoting bonding by mechanical interlocking. In practice, both effects can influence the mechanical properties of a joint, but which will be the most dominant often cannot be predicted. Unfortunately, only a few studies of this subject have been reported, but some examples for silicon nitride can be cited.

The effects of grinding and polishing the bond faces in silicon nitride prior to brazing with pure aluminium braze were examined using a bend test [34]. The bonding temperature was 800 °C and the time was 10 min; Fig. 10.14 shows the Weibull plots of the strengths achieved. It is clear that the roughening of the bond face made the joints weaker. This is because the roughly ground bond face had a damaged surface layer through which fracture of the joint took place.

The weakening of the aluminium brazed joint by a damaged layer is closely related to the water-induced stress corrosion effect of a ceramic. Thus, the strengths of silicon nitride and the aluminium brazed joint in an inert atmosphere and in water [36] are compared in Fig. 10.15. The

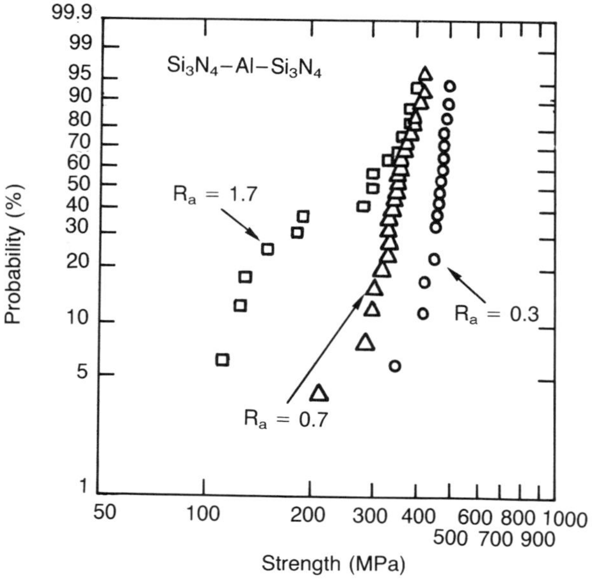

Fig. 10.14 Weibull plots of the three-point bend strength of a silicon nitride joint with bond faces ground using different conditions and brazed with aluminium at 800 °C for 7 min in argon. R_a is the average surface roughness in microns.

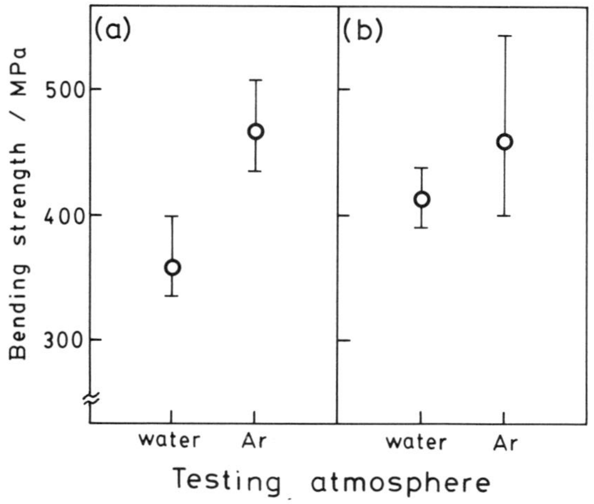

Fig. 10.15 Bend strengths of silicon nitride joints without any interlayer as a function of joining temperature. The joining pressures were 20 MPa (solid symbols) and 1 MPa (open symbols). The sintering additives were (a) $MgO–Al_2O_3$ and (b) $Y_2O_3–Al_2O_3$.

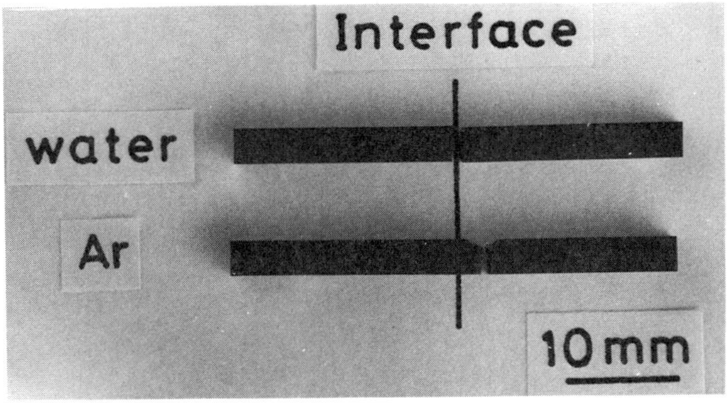

Fig. 10.16 Comparison of the fracture paths of a silicon nitride joint brazed with aluminium at 850 °C for 15 min in a vacuum in different testing atmospheres. The joint fractured completely along the interface in water but did not in argon.

strength of silicon nitride was decreased from 470 MPa to 410 MPa by changing the test atmosphere from argon to water. However, the strength of the joint which had rough bond faces decreased from 470 to 360 MPa. Figure 10.16 shows the fracture paths of the joint in both atmospheres: the fracture passes through the damaged layer and along the interface in water, but does not in argon. Thus a rough bond face joint could have a strength as high as that of silicon nitride itself if tested in an inert atmosphere, but stress corrosion of the damaged layer can weaken the joint.

10.5 JOINING OF SILICON NITRIDE TO ITSELF

Joining silicon nitride to itself can be performed using metallic interlayers as described previously [35]. The main advantage in using metallic materials is that it becomes possible to decrease the fabrication temperature, pressure and time. However, they cannot give both high-temperature strength and oxidation resistance to a joint. Therefore direct joining and joining using oxide interlayers are discussed in this section.

Solid state joining of two pressureless sintered silicon nitrides with different sintering additives ($MgO-Al_2O_3$ and $Y_2O_3-Al_2O_3$) has been evaluated using a nitrogen atmosphere at temperatures of 1400–1600 °C with or without applied pressure [37]. Figure 10.17 shows the bend strength as a function of the joining temperature. The joint strengths of $MgO-Al_2O_3$-doped silicon nitride were higher than those of the $Y_2O_3-Al_2O_3$-doped ceramic and reached that of the bulk ceramic when bonded at temperatures above 1500 °C. The $MgO-Al_2O_3$ grain boundary phase promoted deformation of silicon nitride at the joining temperature. The

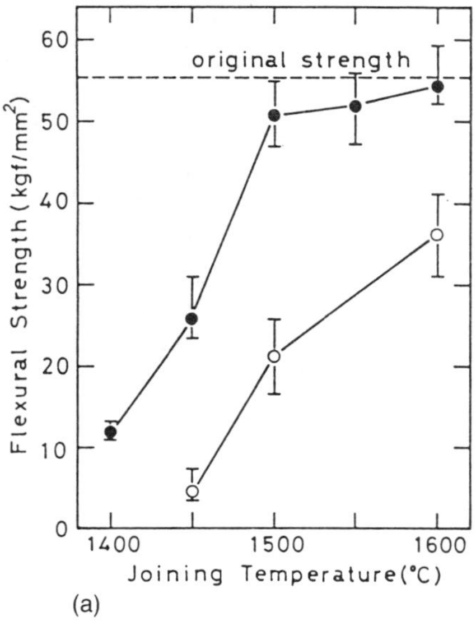

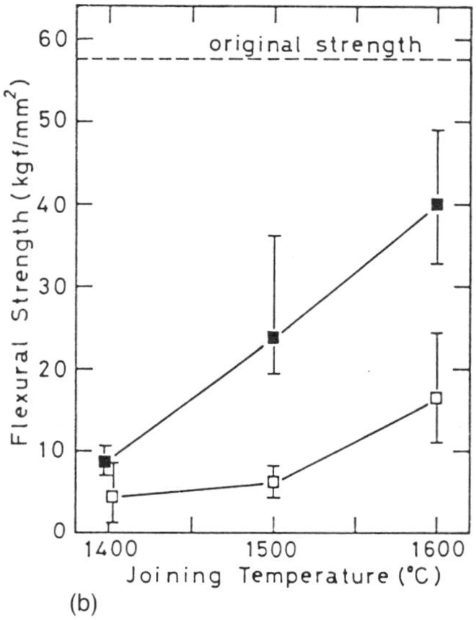

Fig. 10.17 Comparison of the four-point bend strength of (a) a silicon nitride joint brazed with aluminium at 850 °C for 15 min in a vaccum and (b) silicon nitride, in different testing atmospheres.

viscosity of the grain boundary phase was very important because some softness was required to achieve complete contact at the interface at the joining temperature.

Solid state bonding of HP and RB silicon nitride using ZrO_2 or $ZrSiO_4$ has been examined [38]. Bonded high-purity (HP) silicon nitride had a moderate strength, but lowering the silicon nitride purity degraded the strength because of the formation of a ZrN interface layer which affected the oxidation of silicon nitride.

An oxide glass can also be used for joining HP silicon nitride [39–41]. The glass composition was chosen to approximate the oxide grain boundary phase in silicon nitride, e.g. $MgO-Al_2O_3-SiO_2$, and joining was carried out in a nitrogen atmosphere in the temperature range between 1550 and 1650 °C without pressure. The liquid phase reacted with and penetrated the silicon nitride to produce an interfacial phase of Si_2N_2O, β-Si_3N_4 and oxynitride glass. The maximum bending strength of the joint was above 450 MPa when the bonding layer was thin (about 30 μm).

10.6 SUMMARY

Particular attention must be paid to two effects when joining silicon nitride to metallic materials: interfacial bonding and residual stress. A wide variety of joining methods that depend on the fact that silicon nitride reacts with many metallic elements to form well-bonded interfaces have been developed for bonding silicon nitride to metals. Several categories can be involved and each process has its own advantages. However, much remains to be learnt. Thus further experimental evidence could allow a more detailed understanding of the stress corrosion effects produced not only by water but by other corrosive and oxidizing environments as well. Small components (up to 15 mm in diameter) can be joined using soft metal or laminated interlayers. How to join a large-size silicon nitride component to a metal, however, is still a major problem because of the severe residual stresses. Since most silicon nitride components will be used at elevated temperatures, evaluation of high-temperature properties such as strength, oxidation and thermal stress is required.

REFERENCES

1. Katayama, K., Watanbe, T., Matoba, M. and Katoh, N. (1986) *SAE Tech. Pap. Ser. 861128*.
2. Voitovich, L.F. (1971) *Refractory Compounds*, Naukovy Dumuka, Kiev pp. 96–103.
3. Siebels, J.E. (1983) in *Progress in Nitrogen Ceramics* (ed. F.L. Riley), Martinus Nijhoff, pp. 455–63.
4. Mizuhara, H. and Mally, K. (1985) *Weld. J.*, **64**, 27–32.
5. Ramsey, M.J. and Lewis, M.H. (1985) *Mater. Sci. Eng.* **71**, 113–22.
6. Ishida, Y., Ichinose, H. and Tanaka, S. (1985) Presented at *SEIKEN Symp. on Interface Structure, Chemistry, Mechanical Properties and Diffusion Bonding, Tokyo*, 2–4 December 1985.
7. Johnson, S.M. (1987) Presented at *Int. Forum on Joining of Structural Ceramics*, Pittsburgh, PA, 29–30 April 1987.
8. Suganuma, K., Okamoto, T., Koizumi, M. and Shimada, M. (1985) *J. Mater. Sci. Lett.*, **4**, 648–50.
9. Kohno, A., Yamada, T. and Yokoi, K. (1985) *J. Jpn. Inst. Met.* **49**, 876–83.
10. Suganuma, K., Okamoto, T., Koizumi, M. and Shimada, M. (1987) *J. Mater. Sci.*, **22**, 1359–64.
11. Ning, X.S., Suganuma, K., Morita, M. and Okamoto, T. (1987) *Philos. Mag. Lett.*, **55**, 93–7.
12. Naka, M., Mori, H., Kobo, M., Okamoto, I. and Fujita, H. (1986) *J. Mater. Sci. Lett.*, **5**, 696–8.
13. Müller, K. and Rebsch, H. (1966) *Silikattechnik*, **17**, 279–282.
14. Morita, M., Suganuma, K. and Okamoto, T. (1987) *J. Mater. Sci. Lett.*, **6**, 474–6.
15. Sata, T. and Urano, T. (1970) *Yogyo-Kyokai-Shi*, **78**, 21–9.
16. Morozumi, S., Kikuchi, M. and Sugai, S. (1981) *J. Jpn. Inst. Met.* **45**, 184–9.
17. Suganuma, K., Okamoto, T., Miyamoto, Y., Shimada, M. and Koizumi, M. (1986) *Mater. Sci. Technol.*, **2**, 1156–61.
18. Suganuma, K., Okamoto, T. and Shimada, M. (1984) *High Temp. High Press.*, **16**, 627–35.
19. Koizumi, M., Takagi, T., Suganuma, K., Miyamoto, Y. and Okamoto, T. (1986) Presented at *World Congr. on High Technology Ceramics, VI, CIMTEC, Milan, 23–27 June 1986.*

20. Tennenhouse, G.J., Ezis, A. and Runkle, F.D. (1985) *Commun. Am. Ceram. Soc.*, **68**, 30–1.
21. Suganuma, K., Okamoto, T., Koizumi, M. and Shimada, M. (1985) *Commun. Am. Ceram. Soc.*, **68**, 334–5.
22. Suganuma, K., Okamoto, T. and Kamachi, K. (1987) *J. Mater. Sci.*, **22**, 2702–6.
23. Suganuma, K., Okamoto T., Koizumi, M. and Shimada, M. (1986) *Am. Ceram. Soc. Bull.*, **65**, 1060–4.
24. Klomp, J.T. (1971) *Weld. J.*, **50**, 88–90.
25. Nicholas, M.G. and Crispin, R.M. (1982) *J. Mater. Sci.*, **17**, 3347–60.
26. Ebata, M. and Koze, S. (1983) *Nikkei Mech.* **6**(20), 185–6.
27. Nomura, O., Ebata, Y. and Hijikata, K. (1982) *Yogyo-Kyokai-Shi*, **90**, 658–63.
28. Pattee, H.E. (1972) *Bull. Ser. Weld. Res. Coun.*, **178**, 1–43.
29. Suganuma, K., Shimada, M., Okamoto, T. and Koizumi, M. (1983) *Commun. Am. Ceram. Soc.*, **66**, 117–8.
30. Suganuma, K., Okamoto, T., Shimada, M. and Koizumi, M. *Commun. Am. Ceram. Soc.*, **67**, 256–7.
31. Elssner, G. (1987) M.P.I., Stuttgart personal communication.
32. Kohno, A. and Yamada, T. (1988) personal communication.
33. Suganuma, K. (1987), unpublished work.
34. Yamada, T., Ekiguchi, H., Okamoto, H., Azuma, S., Kitamura A. and Fukaya, K. (1986) *2nd Int. Symp. on Ceramic Materials and Components for Engines, April 1986*, D.K.G., Bad Honef, FRG.
35. Suganuma, K., Okamoto, T., Koizumi, M. and Shimada, M. (1986) *Adv. Ceram. Mater.*, **1**, 356–60.
36. Suganuma, K., Morita, M., Okamoto, T. and Koizumi, M. (1988) *J. Adhesion Sci. Technol.* **2**, 43–57.
37. Tabata, H., Kanzaki, S. and Nakamura, M. (1984) *Proc. Int. Symp. on Ceramic Components for Engines*, pp. 387–93.
38. Becher, P.F. and Halen, S.A. (1979) *Am. Ceram. Soc. Bull.*, **58**, 582–6.
39. Loehman, R.E. (1981) in *Surface and Interfaces in Ceramic and Ceramic–Metal Systems* (eds J.A. Pask and A.G. Evans), p. 701, Plenum, New York.
40. Johnson, S.M. and Rowcliffe, D.R. (1985) *J. Am. Ceram. Soc.*, **68**, 468–72.
41. McCartney, M.L., Sinclair, R. and Loehman, R.E. (1985) *J. Am. Ceram. Soc.*, **68**, 472–8.

11

Ceramic–metal diffusion bonding – a case history

T. Joy

11.1 INTRODUCTION

The Daresbury Laboratory of the Science and Engineering Research Council operates a 20 million volt Van de Graaff generator providing the high voltages to accelerate a wide mass range of ion beams for research in nuclear structure physics. The accelerator has been operated for 24 h a day since the end of 1982. During this period the Nuclear Structure Facility has become internationally recognized, providing a world-class accelerator serving the UK nuclear physics community and several research groups from overseas. The accelerator itself has been fully described in the specialist literature [1] and its main features are shown in Fig. 11.1.

A critical component of the accelerator is the evacuated tube which carries the accelerated ion beam through the insulating column supporting the high-voltage terminal. The high-voltage aspects of the development of the tube and its operational performance have been described in two review papers [2,3], but little has been reported about the programme of research and development of the metal–ceramic bonding technology which was required to achieve the objectives established in the original design study for the Nuclear Structure Facility [4].

11.2 DESIGN OBJECTIVES

The accelerator tube was to be supported inside the insulating column of the high-voltage generator which is enclosed inside a steel pressure vessel 45 m long × 8 m diameter. The high-voltage structure was designed to be capable of holding off 30 million volts and to use SF_6 gas at a pressure of 8 bar as the insulating medium. To accelerate heavy ion beams, a prime design requirement for the Facility is a very clean high vacuum (better than 10^{-7} mbar) with very low partial pressures of hydrocarbon molecules.

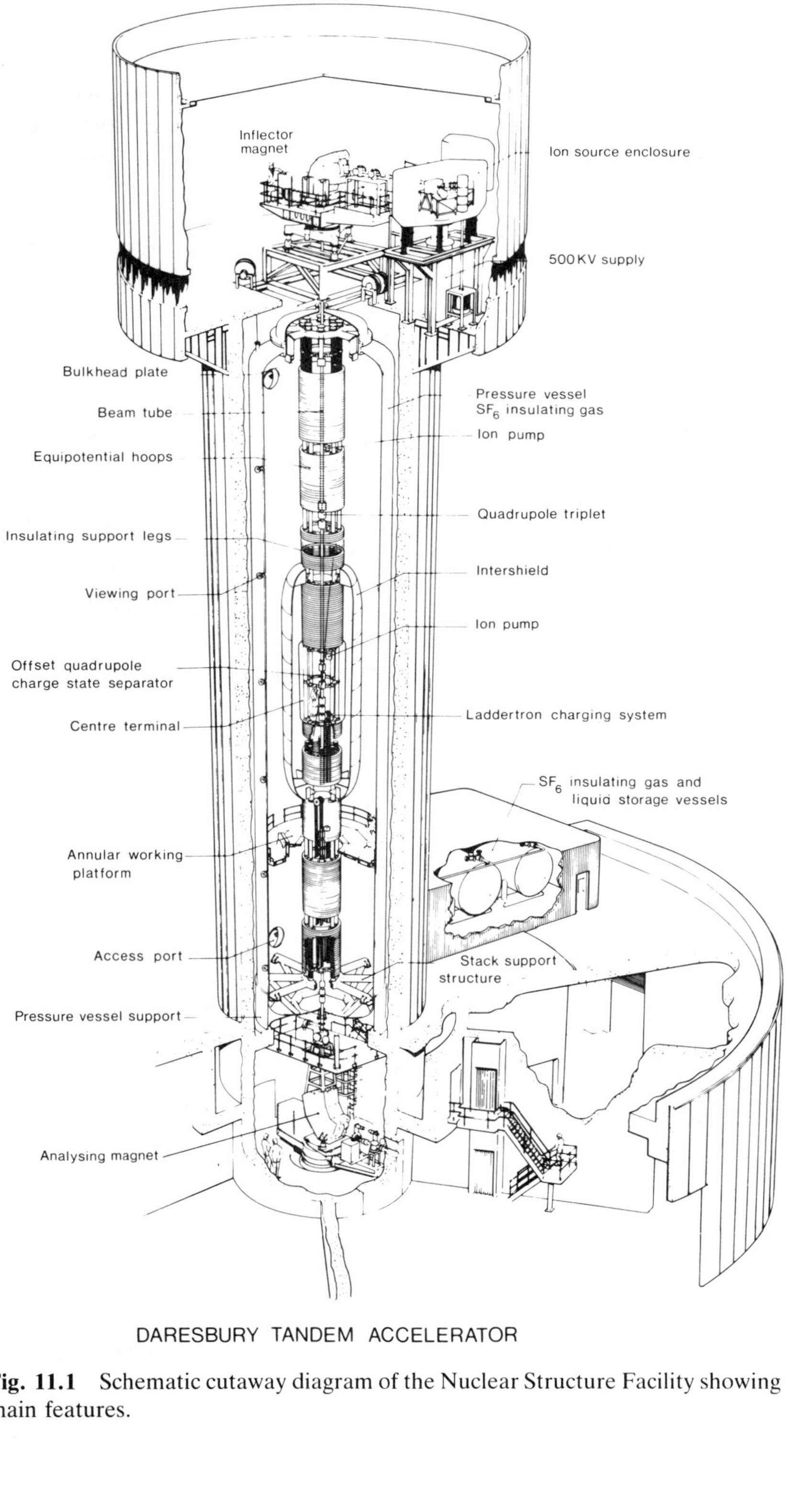

DARESBURY TANDEM ACCELERATOR

Fig. 11.1 Schematic cutaway diagram of the Nuclear Structure Facility showing its main features.

Mechanically, the tube had to withstand repeated cycling between external pressures ranging from 1 bar to 8 bar during transfers to and from storage of the 120 tonnes of SF_6 gas when the tank was opened for maintenance work.

An important objective was that the tube had to withstand not only a high d.c. voltage gradient of up to 2 MV m^{-1} during normal operation but also sudden high-voltage discharges of the complete generator. These discharges occur from time to time when the accelerator is operating close to its conditioned voltage limit and they generate very fast high-voltage transients which dissipate the stored energy of the tandem (over 10^5 J at 20 MV) very rapidly. Under these conditions effective protection of tube insulators is essential and a method of construction which minimizes the probability of bulk electrical breakdown is important.

Tubes for electrostatic accelerators consist essentially of a sandwich structure of insulating rings and metal field-defining electrodes suitably bonded together to form a continuous evacuated envelope for the beam (Fig. 11.2). The geometrical design of the Daresbury tube was based on studies of the high-voltage performance of tubes already in use in smaller accelerators coupled with detailed ion trajectory calculations and studies of the processes affecting high-voltage breakdown in vacuum [2]. On the basis

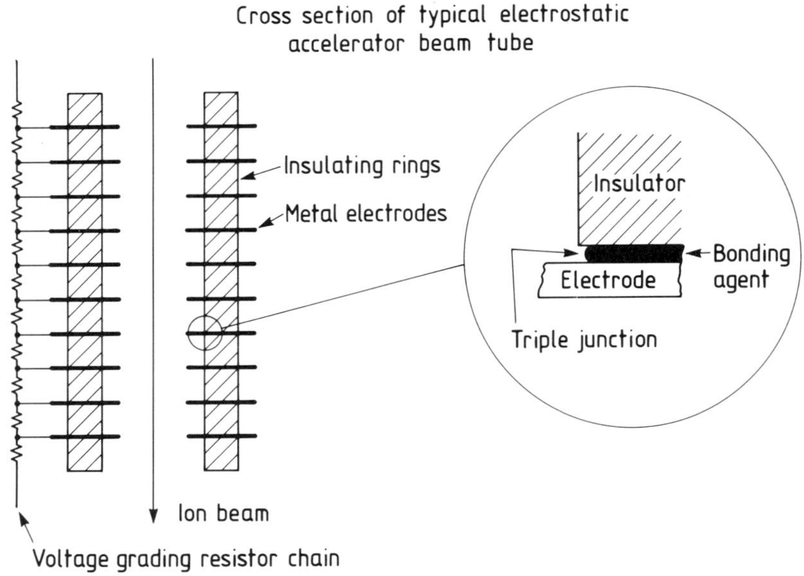

Fig. 11.2 Cross-section of a typical electrostatic accelerator beam tube highlighting the 'triple junction'.

of this work and considerations of vacuum, insulator shielding and pumping requirements the external diameter of the insulating section of the tube was chosen to be 190 mm.

Suitable materials for the construction of the tube were defined by a high-voltage development programme which indicated that the best material for the insulating rings was either borosilicate glass of the highest quality or high-alumina ceramic. Stainless steel or titanium and some of its alloys emerged as the preferred electrode materials. Various established techniques existed for bonding tube sections together including a variety of metallizing/brazing techniques and ceramic or glass frit processes. Organically based adhesives, particularly polyvinyl acetate (PVA), had been used successfully for the manufacture of tubes with less stringent vacuum requirements, but residual gas analyses indicated that their use was incompatible with the clean hydrocarbon-free vacuum conditions which were required. None of the various brazing techniques offered a reliable means of precisely controlling the 'triple junction' between the insulator, the electrode and the 'glue' (Fig. 11.2). The high-voltage breakdown properties of an insulator surface in vacuum are critically dependent on the properties of this junction. It is essential that it should not provide a trigger for electrical breakdown or a source of field-emitted electrons from the cathode side. The use of a ceramic insulator makes this control and the achievement of a void-free bonded interface especially important. The dielectric constant of ceramics with good high-voltage properties is about 9, which means that electric field enhancements due to voids or triple-junction gaps can be large.

What was required therefore was a non-organic joint compatible with good high-voltage performance and with the materials chosen for the construction of the tube.

Very little information could be found in the literature at the time on the technology of direct bonding of metals and insulators, but there were some references to the achievement of successful bonds between various metals and ceramics [5,6]. Progress had also been made in the commercial production of accelerator tubes of a much smaller diameter than had been envisaged for the Daresbury machine by this kind of technique, although no details had been published [7]. Solid phase bonding between dissimilar materials involves either a direct process or the use of an intermediate layer of another material, usually a soft metal or alloy which can be directly bonded to both components of the joint. A general review of early developments in the understanding of this process has since been published [8].

The available information indicated that a directly bonded joint could be achieved for titanium and stainless steel to ceramic by means of an aluminium interlayer, but at that time there was no guarantee that a production process could be developed to produce a tube with the requisite geometry, high-voltage properties and high-vacuum integrity. A literature search indicated that there was very little experience to draw on concerning direct bonding of metals and glasses. Although the use of glass was attractive from the point of view of its superior bulk electrical breakdown properties, the prospect of developing a production bonding process was judged to be remote. However, the prospect of success for a direct metal–ceramic bond was judged to be high enough to warrant the setting up of a collaboration with the Harwell Laboratory to undertake a feasibility study and, in particular, to establish the conditions under which strong vacuum-tight bonding could be achieved between high-alumina ceramic and either stainless steel or titanium and its alloys.

The choice of ceramic was not only based on tests of d.c. and impulse voltage characteristics of a variety of commercially available materials but also required the selection of a ceramic which might be suitable for initial bonding trials. Without detailed knowledge of the factors affecting the quality of the bonded interface, a ceramic was selected which had a grain structure and glassy phase composition which gave good adhesion and wetting in brazed joints. It also had to be capable of being precision ground since the tolerances required to withstand even modest pressures during the bonding operation could not be obtained with as-fired material.

Initial development efforts were concentrated on establishing the required bonding conditions between high-alumina ceramic (96.7% Al_2O_3, Wade, Ireland, UL300) and commercially pure titanium or its alloy 318A which contains 6% Al and 4% V. These materials were all capable of meeting the high-voltage design requirements for insulators and electrodes.

Preliminary investigations quickly established that a direct-bonding process is particularly well suited for the joining of titanium and alumina because of the close similarity of their thermal expansion coefficients. Indeed, the matching of thermal expansion coefficients not only at ambient temperature but throughout the temperature range required to establish a bonded joint proved to be a crucial parameter in determining whether a bonded joint would 'self-destruct' as a result of residual stress after cooling down from the pressing temperature.

Subsequent investigations demonstrated that there is an inverse relationship between bonding strength for metal–alumina joints via an aluminium interlayer and the thermal expansion coefficient of the metal component of the joint [9]. The key role played by the thermal expansion coefficients of the two components of the joint makes the bonding of large-diameter rings particularly difficult. For example, sufficiently strong bonding can be achieved between stainless steel and alumina for small test couples where the residual stress is insufficient to destroy the joint. However, on the scale needed for the Daresbury tube the stresses set up can be high enough to shear the ceramic itself as the joint cools. On a small scale it is possible to stress-relieve the joint, but the time-scale required unnecessarily complicates a large-scale production programme.

11.3 PRIMARY BOND REQUIREMENTS

The basic specifications to be met were as follows:

1. high mechanical bond strength (> 15 MPa);
2. high vacuum integrity with a leak rate per joint better than 10^{-8} mbar ls^{-1} (in fact subsequent trials have shown that leak rates better than 10^{-10} mbar ls^{-1} at a 10 bar pressure differential can be achieved for helium);
3. a sound interface between the bonding material and the insulator with freedom from microscopic voids and high spots (both these features adversely affect the voltage hold-off capability of an insulator–electrode joint);
4. precise control of the 'triple junction' for good high-voltage performance of the insulator surface in vacuum;
5. the process must not deform the tube electrodes sufficiently to affect the ion beam trajectory (this problem is most critical at the injection end of the accelerator where beam energies are lowest; in practice the restrictions implied that distortions of individual electrodes must be contained below 0.3 mm and be randomly distributed);
6. the bonding process must not degrade the high voltage properties of the electrode surfaces where the microscopic finish, cleanliness and degree of oxidation had all been demonstrated to affect their high-voltage performance.

11.4 MAIN STEPS IN THE DEVELOPMENT PROGRAMME

Figure 11.3 summarizes the steps leading up to the final production run in the form of a flowsheet. The important elements of the programme were as follows:

1. establishing basic bonding conditions, key parameters and the overall feasibility of achieving the bond requirements;
2. a detailed study of the dominant process parameters of time-scale, pressure required, temperature and intermediate layer thickness using bond tensile strength as a guide. Bond strength alone is not a reliable guide to the vacuum quality of a joint. However, the alternative of using the degree of interfacial contact as a guide to the bond quality proved to be inordinately time consuming.

Most of these trials were performed using standard test couples (Fig. 11.4).

3. a parallel investigation of the problems of scaling the small-scale results into a production-scale operation; this also involved the design and construction of suitable production equipment;
4. final refinement of process parameters and an investigation of the vacuum characteristics of bonded joints;

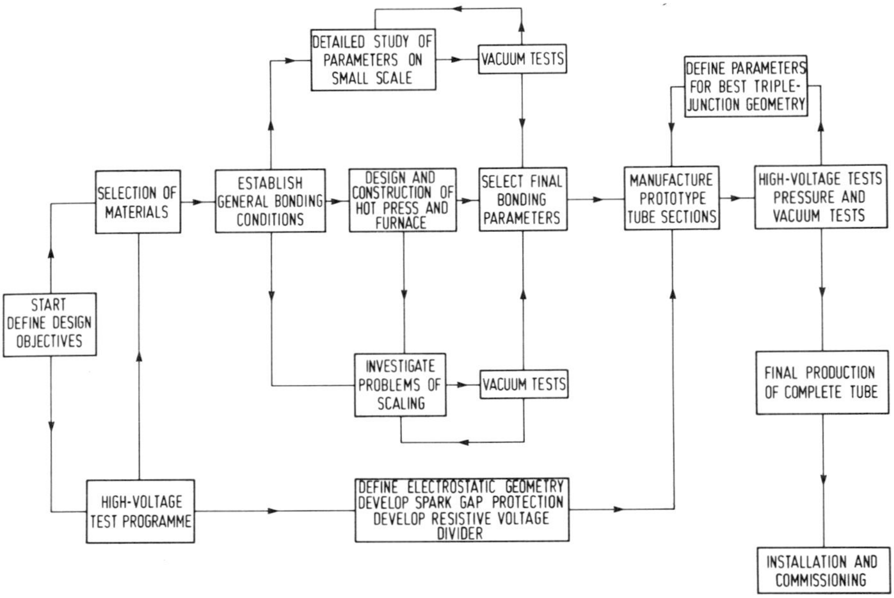

Fig. 11.3 Flowsheet of the accelerator tube production and development programme.

5. production and testing of prototype tube sections, including testing their high-voltage performance;
6. production, installation and commissioning of the MK I tube in the tandem itself.

11.5 PROCESS DEVELOPMENT

Initial trials rapidly established the feasibility of achieving adequately strong bonding between high-alumina ceramic and titanium using aluminium interlayers. In fact it proved possible to produce joints strong enough to fracture within the body of the ceramic during tensile tests. The key parameters identified from small-scale trials were as follows:

1. interlayer thickness: the degree of deformation of the intermediate aluminium layer showed a strong correlation with bond strength. It was clearly a prerequisite for achieving good bonding to expose the aluminium through its oxide surface via the deformation of the foil under pressure;
2. bonding time, temperature and pressure;
3. surface preparation, cleanliness and finish;
4. component geometrical tolerances, particularly of the ceramics, had to

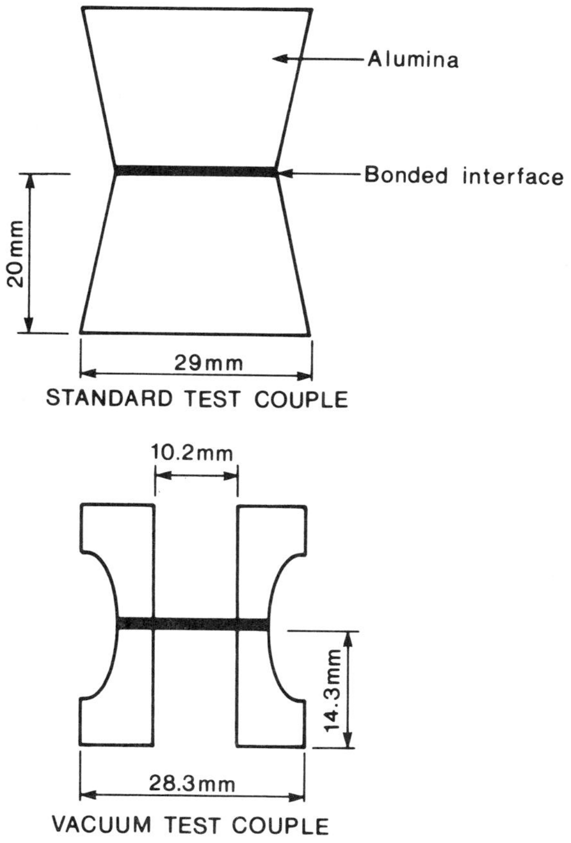

Fig. 11.4 Test pieces used for small-scale investigations of bonding parameters. The pieces with tubular cross-section were used to relate the vacuum tightness of a bonded joint to mechanical strength measurements.

be closely controlled. For example, to obtain the flatness required of the titanium sheet a special process of vacuum creep flattening was developed;

5. Ambient atmosphere in which the pressings are done.

Most of the early trials were carried out by bonding in an inert atmosphere of high-purity argon. Titanium and especially ceramics outgas prolifically at elevated temperatures. The gases involved are mainly water vapour and hydrogen for titanium and water vapour for ceramic. This means that to maintain reasonable purity levels the argon gas environment used for the initial trials had to be continuously circulated and dried.

Although the consistency of the initial results was poor, broad regions of temperature, pressure and time were established where satisfactory bonds of good mechanical strength could be obtained. For example, for pure titanium, bonding could be achieved for a range of intermediate aluminium layer thicknesses up to 1 mm for the following broad range of parameters: time, > 0.5 h; temperature, 400–660 °C (the melting point of aluminium); pressure, 20–60 MPa at the fabrication temperature.

11.6 INITIAL PROBLEMS

Hot-pressing in an inert pure argon atmosphere was initially selected on general cost grounds. However, the chemical reactivity of titanium at elevated temperatures was sufficient to cause oxidation at purity levels of about 1 ppm residual oxygen and nitrogen with the highest-purity argon commercially available. Even this level of purity proved to be difficult to achieve in a press and furnace capable of bonding the full-size tube components. Nevertheless this level was achieved in a working volume of 1 m^3 at 600 °C but was still insufficient to prevent adverse electrical performance resulting from uncontrolled oxidation. There was also an effect on the bond strengths where inconsistent results were obtained in full-scale tests. It became clear that a better prospect for a furnace and press of the scale required was to design a high-temperature press operating under high-vacuum conditions. Such a press was eventually developed to operate at pressures below 10^{-5} mbar at 650 °C, equivalent to an inert gas purity of about 0.01 vol. ppm. After initial oil-free rough pumping, the press was turbomolecularly pumped continuously during the high-temperature cycle ensuring that evolved gases, mainly water vapour adsorbed by the ceramic rings, was more effectively removed than would have been possible with a circulating inert gas stream. The development of a hot-press able to reach 650 °C at pressures below 10^{-5} mbar with low hydrocarbon contamination, 60 000 kg operating load and good mechanical stability eventually turned out to be a crucial factor for the consistent production of bonded full-scale tube sections.

A second problem occurred after establishing the conditions for acceptable bonding for pure titanium. This came to light during transfer of the technology over to the production of larger-size test samples where the deformation of the titanium electrodes after hot-pressing proved to be uncomfortably close to the limits required by the ion optical properties of the tube. As a result of this the conditions required to bond 318A titanium alloy were also investigated as a possible alternative. This material, although very similar to titanium in electrical characteristics and general mechanical qualities, is much more resistant to deformation at elevated temperatures. Bonding can be achieved for very similar conditions to those needed for pure titanium, although higher temperatures are required to reach the same bond strengths.

Although this provided a useful alternative, and indeed 318A alloy has been used to produce tube since the original production run, careful optimization of parameters made it possible to produce the complete MK I tube from pure titanium.

The tube was intended to be modular in construction, made up of individual short sections for ease of assembly and future maintenance. Two further problems immediately became apparent when attempts were made to produce bonded stacks of alternate ceramic rings and titanium electrodes.

1. The first problem concerned the treatment of the ends of the stack where, although the expansion coefficients of the components are similar, there is still sufficient mismatch at the bonding temperature to drive the ceramic rings into strong radial compression when the stack has cooled to ambient temperature. This frequently resulted in failure of the end-joint if the stack was terminated with a metal electrode as had been the original intention. The solution required the use of two electrically redundant ceramic rings at either end of the stack to balance the assembly thermally at the end-joint.
2. The second problem, which arose as a result of attempting to bond multiple stacks of ceramic rings, was that even with the closest control of tolerances the tendency to brittle-fracture under tensile stress makes simultaneous pressing of a large number of ceramic rings very difficult. From the point of view of high-voltage performance it is important to maximize the effective insulating length of the tube modules. This implies that the number of flanged joints involved should be minimized. The longer the effective insulating length is, the lower is the overall voltage gradient on the tube. However, accumulating tolerances imposed practical limitations on the length of bonded assemblies which could be produced with an acceptable success rate. Figure 11.5 summarizes the results of attempting to bond multiple assemblies where the unit bond conditions were close to optimum. Although the data for the larger assemblies are based on trials with a small number of samples, they do illustrate very well the complex interplay of the geometrical tolerances of components and the mechanical limitations of the hot-pressing equipment for the successful bonding of multiple metal–ceramic assemblies. These problems could probably be overcome, but at the expense of, for example, more costly machining processes and more massive pressing equipment.

The central problem of the development programme was to translate the satisfactory results of small-scale tests into a full-scale production process using relatively expensive materials where, for example, a leak of 10^{-10} mbar l s^{-1} in one interface of a multiply bonded stack could lead to its rejection during final tests.

For the most consistent results it was found to be vital to control the

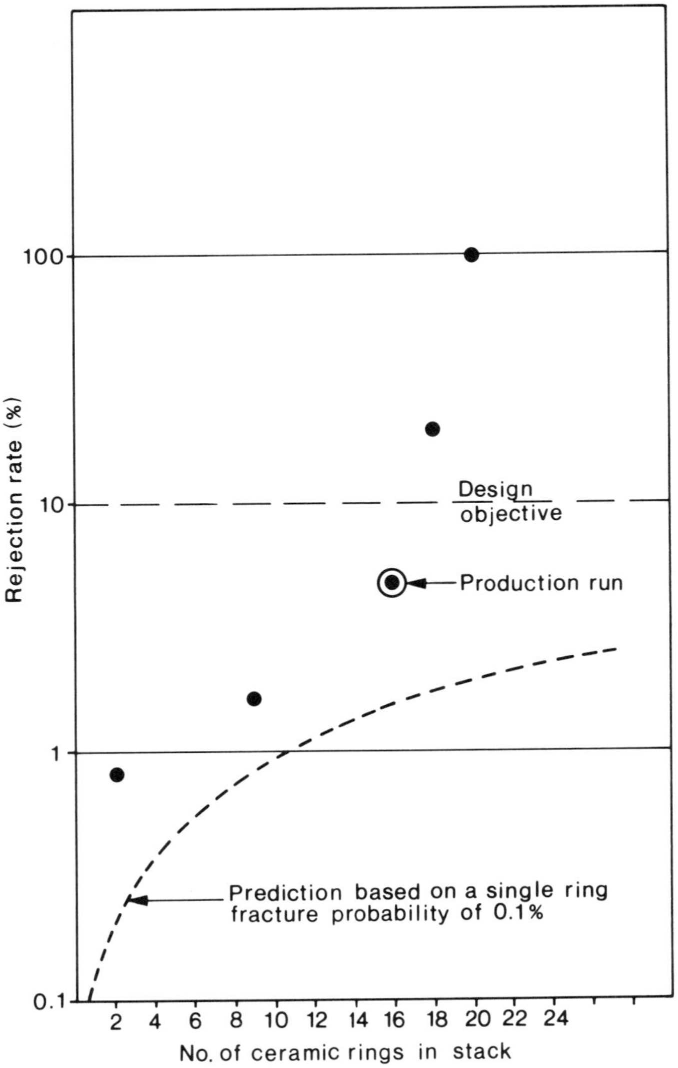

Fig. 11.5 Rejection rate as a function of the number of ceramic rings in the bonded stack.

surface cleanliness and preparation of the components. The presence of significant hydrocarbon contamination arising from inadequate cleaning of the ceramics, for example, tended to inhibit bond formation. An essential part of the programme to minimize the rejection rate was to set up rigid quality control and inspection procedures at each stage of manufacture and preparation of the components.

11.7 THE MK I TUBE

The final design of the tube to be installed in the accelerator involved the pressing of stacks of 16 alumina–titanium sandwiches. Each stack contained $15 \times 4 = 60$ bonded interfaces and 144 stacks were required for the manufacture of the submodular units of the complete tube. Before the final production run was undertaken, a detailed examination of three key parameters was made using complete assemblies of 16 sandwiches. The objective was to achieve for the production run a reject rate of less than 10% for the tube submodule assembles. As Fig. 11.5 indicates, this looked feasible but difficult.

The complete tube required 36×4 submodular assemblies plus spares with one in four units having an integral titanium bellows section welded to the end-flange to facilitate assembly and possible future replacement. The whole process was reviewed at this stage from the following standpoints:

1. effect of the variation of the thickness of the intermediate bonding foil on the control of the triple junction geometry;
2. control of geometrical tolerances and final electrode deformation;
3. the surface finishes, tolerances and preparation of components and their effect on high-voltage and vacuum performance.

These full-scale pre-production trials established, for example, that control of the ceramic machining tolerances, particularly flatness, and the microscopic finish had to be very stringent. The flatness and parallelism required for the ceramic rings, for example, was specified to be within 0.01 mm or better across the 190 mm diameter ring of the MK I tube insulator. It was discovered that not only was the general microfinish achieved during grinding of the bonding surfaces of the ceramic important, but also the macrofinish played an important role. The macrofinish is due to imperfections and tolerances in the grinding equipment, for example, which override the finish imposed by the fineness of the diamond grinding head. The importance of the control of this aspect of the surface finish has also been emphasized in theoretical considerations of metal–metal bonding conditions [10].

11.8 THE PRODUCTION RUN

Table 11.1 lists the parameters for the final production run for the MK I tube together with geometrical and production details. The production schedule to produce 144 submodules of tube required the use of three hot-presses working continuously to make six bonded assemblies per week.

Initial pumping of the furnace was via oil-free diaphragm roughing pumps to pressures below 10^{-1} mbar. High vacuum was then established by means of a $450\ \text{l s}^{-1}$ turbomolecular pump reaching 3×10^{-5} mbar in

Table 11.1 Tube and production process parameters

Submodule length	196 mm
Insulator	16 per submodule, two of these for thermal balancing Material 96.7% isostatically pressed alumina, type UL300 (Wade, Ireland) Diamond ground on all faces Surface finish 1 μm R_a Thickness, 10.00 mm Outside diameter, 190.00 mm Inside diameter, 150.00 mm
Electrodes	Flat commercially pure titanium Thickness, 1.00 mm Outside diameter, 203 mm Inside diameter, 112 mm Internal edge radiused
Inserts	Pure titanium, polished and vacuum baked
Construction	High-temperature diffusion bonded using 0.1 mm Al interlayer; pressed at 550 °C for 1 h at 56 MPa
Vacuum integrity of submodule	$< 10^{-10}$ mbar l^{-1}s at 10 bar He pressure
Production rejection rate at this leak rate	5% of submodules
Connecting flanges	Pure titanium
Seals	Pure aluminium (Daresbury design)

about 1 h. The temperature distribution inside the press was monitored by Pt–Rh thermocouples which automatically provided servocontrol of the hot zone to give a temperature uniformity over the pressing of ± 2.5 °C at up to 650 °C. The pressure inside the press at the peak of the temperature cycle is less than 7×10^{-6} mbar. A disadvantage is the long cycle time of about 24 h which is due primarily to the massive bulk of the press and the need to minimize the possibility of thermal shock or stresses.

Each bonded submodule underwent an initial vacuum test before fitting out with titanium end-flanges. Subsequently a final high-vacuum test at high pressure (10 bar) was carried out for each submodule using residual gas analysis equipment capable of detecting leaks of 10^{-10} mbar l s^{-1} for helium. This extremely searching test had to be passed successfully by each assembly before final fitting out with its external resistive divider network

Fig. 11.6 A submodule 16 ceramics high in its alignment jigs ready for pressing.

and protective spark-gap assemblies which protect the insulator against damage caused by electrical transients. In addition, all completed submodules of the tube were given a high-voltage test. This rigorous test schedule ensured that pressures of better than 10^{-7} mbar could be

Fig. 11.7 A general view of the pressing laboratory.

maintained throughout the 40 m long beam path of the complete tube installed in the accelerator.

Figure 11.6 shows an assembly 16 ceramics high in its alignment jigs ready for pressing and Fig. 11.7 shows a general view of the pressing

Fig. 11.8 Completed tube submodules.

Fig. 11.9 Tube submodule without field defining inserts.

Fig. 11.10 A tube module installed inside the accelerator column.

laboratory. Figures 11.8 and 11.9 show completed submodules of tube and Fig. 11.10 shows the tube installed inside the tandem column.

11.9 CONCLUSIONS

This case history in metal–ceramic fabrication has described the application of solid phase bonding technology to the production of the MK I tube for the Nuclear Structure Facility at Daresbury.

With a few exceptions, the performance of Van de Graaff accelerators is limited by the tube. However, the MK I tube has enabled the Facility to achieve its initial design operating voltage of 20 million volts with excellent transmission and good optical quality of heavy ion beams. Although occasional damage resulting from the effects of high-voltage transients has occurred, no failures of the metal–ceramic interfaces have been encountered either during commissioning or in continuous operation of the accelerator for scheduled nuclear physics experiments. Indeed, some of the original tube sections which have recently been replaced by an improved design have shown no deterioration in the vacuum integrity or the strength of the bonded interface.

It is intended to publish the detailed results and data obtained during this development programme in the near future [11].

ACKNOWLEDGEMENTS

It is a pleasure to acknowledge the fruitful collaboration with my colleagues M.G. Nicholas and R. Crispin at the Harwell Laboratory. The successful development of the full-scale production process at Daresbury owes much to the meticulous technical contribution of J. Iliffe, with G. Corrie and J.M. Leese providing the high standard of engineering support which was essential for the success of the programme.

The close cooperation of C.G. Martin and the staff at Wade, Ireland, throughout this programme is also appreciated.

REFERENCES

1. Aitken, T.W., Halliday, B.S., Johnstone, W.T., Jones, C.W., Joy, T., Morris, M.C., Tait, N.R.S. and Voss, R.G.P. (1974) *Nucl. Instrum. Meth.*, **122**, 235.
2. Joy, T. (1981) *Proc. 3rd Int. Conf. on Electrostatic Accelerator Technology, Oak Ridge, April*, IEEE Publishing Services, New York, USA.
3. Aitken, T.W., Eastham, D.A., Joy, T., Leese, J.M., Tait, N.R.S. and Thorn, R. (1985) Presented at *4th Int. Conf. on the Technology of Electrostatic Accelerators and Associated Boosters, Argentina, April*, North Holland, Amsterdam.
4. *Daresbury Lab. Rep. (1973) DNPL/NSF/R4*, R.G.P. Voss.
5. Klomp, J.K. (1970) *Sci. Ceram.*, **1**, 501.
6. Comsa, G. and Simionescu, C. (1964) *J. Sci. Instrum.*, **41**, 352.
7. Herb, R.G. (1971) *IEEE Trans. Nucl. Sci.*, **18**, 71.
8. Mortimer, D.A. (1979) *AERE Rep. R9533*, AERE Harwell.
9. Nicholas, M.G. (1986) *Brazing Soldering*, **10**, 11.
10. Garmong, G., Paton, N.E. and Argon, A.S. (1975) *Metall. Trans. A*, **6**, 1269.
11. Crispin, R.M., Joy, T. and Nicholas, M.G. in press.

Index

Accelerator
bonding 194, 199, 200, 202
design 196–8
production 205–10
Acoustic
microscopy 161, 162
Activity
chemical 76
coefficient 77
gradient 19
Alumina
aluminium, diffusion bonding 101, 104
copper, diffusion bonding 144
insulating rings 197
metal interface strengths 134, 135
nickel, diffusion bonding 102
niobium, diffusion bonding 100, 103,
139, 141–4
platinum–hydrogen system 119–23
reduction 119, 120
Aluminium
braze for silicon carbide 75
braze for silicon nitride 177, 178
diffusion bonding interlayers 101, 135, 200
solution in platinum 119

Bond quality
of glass-ceramic joints
Brazing
conditions 82
defects 65
glass-ceramics 38, 39
metallized ceramics 65, 66
prior surface preparation 83
thermal cycles 84, 85

Ceramic and Metal Properties
comparison 1, 2
Ceramic chemistry 58
Compliance 104, 151
Chromium
oxidation 117
oxide reduction 118
Cobalt oxide
at glass metal interfaces 22
Contact angle 15–17, 50, 73, 114
Copper
coated with glass ceramic 48
indium-titanium braze 78, 79, 90
oxidation 117
oxide reduction 118
tin–titanium brazing of diamonds 91
titanium alloy, reactivity 80

Diffusion
in glasses
of glassy phases
of plated coatings
Diffusion bonds
accelerator modules 200–10
applications 95, 194–210
fracture mechanics 139–45
interlayers 104, 105, 185–8, 199–210
silicon nitride 178

Direct bonding
  glass ceramics 42–7
Dundurs parameters 149

Energy
  glass surfaces 18
  interface fracture energy 105–110, 144, 148, 149, 128, 129, 133
  interface 16
  liquid surface 16, 73, 114
  metal surface 14, 18, 114
  metal surface 14, 18, 114
  solid surface 16, 73, 114
  solid–liquid interface 73, 114
Environment
  for accelerator bonding 201, 202, 205, 206
  for brazing 83, 84
  for diffusion bonding 102, 201, 202
  for diffusion of coatings 65
  for metallisation 63

Fracture resistance 128, 129, 146

Glass ceramics
  coatings 45, 47–9
  coating application 45
  lithium-alumino-silicate 44
  zinc-alumino-silicate 45
Glass metal seals
  compression 8–11
  housekeeper 8, 9
  manufacture 14, 25–8
  type 8
Glazing
  glass ceramics 39–42
  silicon nitride 191

Hafnium
  silicon nitride joints 151
Hydrogen – water ratios 118–20

Iron
  chromium, bonded to glass-ceramic 44
  chromium, coated with glass-ceramic 48
  nickel, load throughs 43

Interface pores in diffusion bonds 98–100, 115

Kovar
  sealing glasses 26
  silicon nitride joints 184
  workpieces 8, 11

Leak tightness 85, 199, 203, 206
Low valent oxides
  at glass–metal interfaces 51

Mechanical attachment 3, 4
Metallizing coatings
  application 60
  surface preparation 58
  debasing additions 58
  flow chart 57
  glassy phases 56
  metal constituents 59–61
  paint 59, 60
  primary layer 59–63
  secondary layer 63–5
Microstmetural examination 69
Molybdenum
  Glass-ceramic components 45
  'Moly-manganese' coatings 56, 61

Photoacoustic Microscopy 170
Plane strain fracture toughness 105, 133, 147
Plating
  barrel 65
  blisters 63
  salt removal 63
Platinum
  Alumina – hydrogen system 119–23
Preoxidation
  copper 49
  iron-chromium 49
  kovar 27
  metals 19

Reactions
  during diffusion bonding 101, 102
  at glass-metal interfaces 12, 17–26
  redox 15, 19–25
  silicon nitride-titanium 121

sodium silicate-metals 19–23, 25, 26
Reduction
alumina 119, 120
chromia 118
cuprous oxide 118

Silica
diffusion bonded to steel 104
Silicon carbide
aluminium braze 75
diffusion bonding 104
Silicon Nitride
brazing 176–8
composite interlayers 185, 187
diffusion bonding 104
eutectic joining 179, 180
hafnium joints 151
iron joints 179, 180
kovar joints 184
properties 173
reactions with aluminium 177
reactions with metals 175, 176
solid state bonding to metals 178
steel joints 135, 137, 139
thermal expansion 174
titanium reactions 123–6
turbo charger 174
Silver
Copper–titanium brazes 39, 74, 75, 80, 81, 91, 176, 177
Copper–indium–titanium brazes 74, 75, 81
Solubility
titanium in nickel 123, 124
nitrogen in titanium 123, 125
Sodium silicate glass 19–21
Spinel oxides
at diffusion bonded interfaces 103
Spreading 15
Strength
active metal brazed joints 86, 89–92
alumina-aluminium bonds 83
ASTM tests 86, 90, 92, 105, 201
bond strength maxima 88
braze–ceramic interfaces 86–9
ceramic–metal seals 69–70
conventional tests 129–32
conventional test results 134–9
fracture mechanics tests 132, 133, 147–53
fracture mechanics test results 139–45
monolithic materials 2
silicon nitride–metal interfaces 180–2
tensile strengths 52, 69
test speciments 129–31
Stresses
complex stress intensity factor 148
crack tip 148
glass–metal interfaces 53
Residual 12, 13, 52, 53, 104, 182, 184
Surface
cleanliness 58, 59
contamination 113
finish 58, 906, 206
machining 115
roughness 13, 83, 95, 96, 188

Temperature
active metal brazing 84, 85
bonding accelerator modules 202
critical, for diffusion bonding 98
glassannealing 11
glass-ceramic forming 33
glass setting 11
glass softening 11
glass straining 11
glass working 11
melting 2
Thermal expansion
characteristics 11, 34
coefficients 2, 8, 11, 13, 198
cracking 66–69
differences 135
(mis) matches 40, 185–8, 203
Thermal wave interferometry 167, 168
Thermodynamies
equilibrium 116
irreversible 116
Thermography
principles 164
pulsed video 165, 166
Titanium
alloy 318A 198, 202
brazing foils 82

brazes for glass-ceramics 39
carbide 79, 80
coated with glass-ceramics 48
dioxide debasing additions 58
electrodes 197
hydride 82
monoxide 75–8, 81
nickel braze 74, 81
nitride 81, 122
oxides 75, 76
role in brazing oxides 75–8
role in commercial brazes 81
role in welting non-oxides 79–81
silicon nitride reactions 123–6
silver-copper brazes 39, 74, 75, 81, 91, 177
Tungsten
in metallising paints 60
saling glasses 11, 26

Ultrasonic
detection of flaws 161
spectros copy 160, 161
tests 155–62
test applications 159–60
wave characteristics 155, 159

Van der Waals bonding 15, 51

Welding 3
Wetting
by brazes 74–85
by glasses 15–18, 49–51
Work of adhescon 87, 95, 103, 114

X-radiography
characteristics 162, 163
detection of voids 162, 164

Young's equation 17, 73, 114

Zirconia
brazing 81, 91
nickel, diffusion bonding 102, 103
palladium, diffusion bonding 108, 110
platinum, interfaces 120
Zirconium brazing foil 82